THE CUTTING EDGE

MINING IN THE 21ST CENTURY

THE CUTTING EDGE

JAMES B. GOODE

JESSE STUART FOUNDATION
ASHLAND, KENTUCKY
2002

Private collection of James B. Goode

Wedding photograph of Idaho Missouri Patterson and Larkin Andrew Goode March 16, 1905. Ducktown, Tenn.

Private collection of James B. Goode

James Skelt Parsons operating lathe at Wisconsin Steel's International Harvester Mine Machine Shop, 1930s.

DEDICATION

To my Grandfathers Larkin Andrew Goode and James Skelt Parsons who daily stepped into places where no man had been.

Library of Congress Cataloging-in-Publication Data

Goode, James B., 1948-
The cutting edge : mining in the 21st century / James B. Goode.
p. cm.
Includes bibliographical references.
ISBN 0-945084-83-8
1. Coal mines and mining -- United States. I. Title

TN805.A5 G66 2000
622'.334--dc21 00-034806

Book Design By Jim Marsh

Published By:
Jesse Stuart Foundation • P.O. Box 669 • Ashland, KY 41105 • (606) 326-1667 • JSFBOOKS.com

FOREWORD

By Mike Musulin, II
Former President
Kentucky Coal Association

The idea for this volume was conceived during the Kentucky Coal Association's continuing work on behalf of the coal industry to educate the public about the importance of coal to our state and our nation. As part of this ongoing effort, the Association, in conjunction with the Kentucky Coal Council, publishes a *Kentucky Coal Facts* booklet and sponsors a coal education website on the Internet. During research for these educational tools, a lack of educational material about modern coal technology became apparent, and a realization emerged that a remedy had to be invented.

Professor James Goode, a native of Benham and author of *Ancient Sunshine: The Story of Coal*, was contracted to write a book aimed at middle school children that would accurately describe the modern mining industry. Calling upon his extensive research on mining history, Professor Goode assembled the people, materials, and information needed to publish *The Cutting Edge: Mining in the 21st Century*. With input from mining companies, equipment suppliers, electric utilities, government regulators and elected officials, he was able to tell the interesting story of mining's technological advances in the 20th Century and its plans for the 21st Century. This book recognizes that the stereotype of coal as a pick-and-shovel industry has persisted and that the simple fact is that it has not relied on these tools since the early part of this century.

The Cutting Edge discusses coal in terms of the personal equipment and simple machines that miners use to protect themselves and monitor their workplace, describes the complex underground and surface machinery that make the American coal industry one of the most productive workforces on earth, discusses coal transportation and coal utilization, outlines the regulatory aspects of the coal industry, and examines careers in coal mining and what the future may hold.

Donna J. Slone, Photographer

James B. Goode at Evening Star Farm
Lawrenceburg, Ky
November 1999.

Articles in this book discuss the miner's personal equipment and his use of simple machines. What the modern miner wears to work and the sophistication of this safety gear is examined. These writers discuss everything from the expected (a shovel) to the unexpected (a computer keyboard).

The Cutting Edge contains articles and illustrations which describe underground machinery such as continuous miners, longwall cutting machines, roofbolters, loaders, and shuttle cars. There are discussions and illustrations which depict the specialized surface mining machines of today. Numerous illustrations show the enormous size of today's draglines, powered shovels, trucks, and drills.

Coal can be taken to market in many different ways. Articles on coal transportation and preparation describe the differing paths coal may take to its ultimate destination. Coal can and does travel by pipeline, rail, road, and water after it has been washed and prepared for market.

The majority of coal in this country is used to generate electricity, but articles are also included on how coal is used to make transportation fuels, steel, and many other useful by-products.

The modern technology of coal mining is not restricted to just extraction and transportation. Technology is also apparent in the modern techniques used to reclaim surface mined land. Post mining land use issues and environmental concerns are also addressed in this book and some illustrations of reclaimed land are included.

Government regulation plays a large part in coal mining and one discussion is devoted to the permitting process and how the regulatory agencies provide oversight to the coal industry.

Modern technology will provide jobs for future miners, and throughout *The Cutting Edge* the sophisticated machinery that coal employees need to master in order to survey, map and plan coal mining operations is considered.

This book is a tribute to the men and women who serve their fellow man by providing heat, light and power to the nation. It strives to educate our young people about coal and how valuable coal is to our nation. Coal may be an ancient fuel, but it is as modern and relevant as today's computers. Without coal, there would be no computers, no Internet, or, for the sake of the potential readers, no video games. It is a valuable and modern resource which requires advanced technology to be utilized for the good of humankind.

CONTENTS

U.S. Department of the Interior, Washington, D.C.

Dragline shovel operating in western Kentucky.

INTRODUCTION

By James B. Goode

For most people, coal mining always has been a mysterious occupation fraught with danger and intrigue. Black-faced miners wearing strange devices are seen coming and going from dark holes in the earth. What the average person knows about mining usually is associated with these images or those published by the media—coal union strikes or tragedies depicting images of weeping relatives watching the bodies being removed from the mine. But mining has a rich industrial, economic, social, and technological history.

The technological history is, perhaps, the most interesting. In the beginning, coal mining technology was fairly simple. All a miner needed was a pick, shovel, and some method of removing the coal to the outside—usually a crude wheelbarrow or a small wagon which either he or his mule would pull.

Innovations in mining have progressed along two lines. One has been invention which emerged from within the industry and the other has been co-opted from devices which have been fabricated for other industries. These inventions also have followed certain problem solving patterns. Locating the seam, facing the coal, advancing into the seam, extracting the coal, and transporting the coal to a destination have all been problematic. In addition, the safety of those involved in this dangerous occupation has been a concern since its inception.

In the very early days of mining, prospectors located coal seams by observing where coal outcropped near stream beds; they then dug several test holes to determine the uniformity, size, and quality of the coal. The material was either moved using mules or was carried by wheelbarrows. As technology emerged, bulldozers, steam shovels, endloaders, backhoes, and various types of drills were used in prospecting.

Facing the coal (or preparing the site so that the coal can be mined) required that the trees and dirt be removed. The same machinery used for prospecting also was employed here as well as implements used for logging such as axes, hammers, saws, wedges, cant hooks, bark spuds, chain grabs, saws, winches, and other implements.

As the miner advanced into the coal seam, he had to ensure his safety. Keeping the roof from falling on him was an immediate problem. Utilization of the trees he had removed from the site for this purpose was logical. He cut the wood into lengths and wedged them between the top and bottom of the mine as he removed the coal. Later, he would use crude, man-operated drills to drill into the top and install wooden roof bolts to help hold the rock and slate. The wooden roof bolt was a cylinder of wood of about 2 inches in diameter which had slots cut in each end. A small wooden wedge was placed in each slot. When the roof bolt was placed into the hole, the miner would drive the bolt upward into the hole, advancing the bottom and top wedges into the slots which caused the cylinder to expand. This kept the bolt in the hole and helped hold any material which had the potential to fall. Later, a metal version of this bolt was also developed. Roof bolt technology then advanced to the introduction of the conventional metal bolt which had an expansion device mounted on its end similar to one which had raised striations and used epoxy glue, to a cable bolt which can hold extreme weights. Of course, a class of associated machinery to drill and place the bolt also has been created.

Martin County Coal Corporation, Inez, Kentucky

Charles Hilton, Jr. operating remote control device Martin County Coal Corporation's White Cabin Mine.

The miner also faced the problem of conveying fresh air to the area where he worked. Initially, the simple idea of heat rising helped him solve that problem. He sunk a corresponding shaft connected to the work tunnel, built a fire, and let physics take over the task. Later, electric fans were employed to circulate fresh air. Permanent walls made of stone or concrete block or temporary walls called "curtains" made of burlap and, eventually, plastic were used to direct air flow. Test instruments emerged to measure the velocity of air, amount of undesirable gases, and presence of dust suspended in the air.

As extraction and transportation machines became more complicated, there was a necessity to prevent the miner from becoming injured by wheels, gears, pulleys, rollers, and other devices. A whole science of safety and accident prevention was created. Everything from gas monitors, audible alarms and signals to self-rescuers and oxygen apparati became a part of the industry.

Naturally, the technology used to extract the coal developed rapidly. Hand drills and dynamite soon gave way to chainsaw-like machines which undercut the coal and allowed it to break easier as it was blasted to the floor of the mine. Pneumatic and then electric drills were used. The cutting machines were replaced with more complex machines called continuous miners which used sharp bit-like teeth designed to rip the coal from the face. Longwall miners which "walked" on hydraulic legs and looked very much like giant meat slicers that sheared coal on longer faces were devised. A plethora of complicated loading devices began to be used. Machines containing crab-like legs gathered the coal as it fell from the face and pulled it onto devices which would carry it to the outside.

A variety of underground transportation devices were employed and included such machines as electric locomotives, trolley engines, diesel engines, rubber tire shuttle cars, rubber tire scoops, small rail cars, metal shaker conveyors, chain conveyors, rubber belt lines, ratio feeders, smoke rollers, and hundreds of other contraptions. Outside transportation innovations also emerged. Portable belt loaders, rubber tire end loaders, tipples, railhead loading facilities, unit trains, transport trucks, coal barges, and coal pipelines were among the most prominent.

As surface mining emerged, equally unique technology was also developed to meet the special needs of this industry. Mining the coal, as well as removing, storing, and restoring overburden (rock and dirt), presented special problems. Specialized extraction, loading, and transportation equipment were derived. Augers, drills, blasting devices, large capacity trucks, giant bulldozers, backhoes, endloaders, road grading machines, draglines, steam and electric powered shovels, and other devices were introduced. Reclaiming surface mined land has also spawned new technology. Hydroseeders, tree planters, and associated equipment have been developed.

Coal mining has evolved into a complex, highly technical industry. Computers and printed circuits have become common. Surveying is done using devices which receive information via satellites. Driverless transport trucks now take their commands from computers and go about being loaded and unloaded without humans on board. Coal is being transported in pipelines using liquids which have been treated to reduce friction. New extraction, loading, safety, and transport devices are constantly being developed. Both the government and private manufacturers have active research departments that work incessantly to solve both old and new problems in mining.

Mining in the 21st century and beyond will hold some fascinating discoveries and foster unique innovations. Many believe that when mineral resources are depleted on Earth humans will begin to look to other planets and celestial objects as sources. This will spawn a whole new series of problems to solve. Imagine—jet powered shuttle cars, coal trains of space shuttles linked together and pulled by nuclear fusion powered engines, remote control extraction devices which are operated by humans who sit on Earth in front of computer monitors and slide their fingers across touch pads to operate the controls of a machine located on Asteroid Mining Station Alpha Omega #2000, Sector 7...Imagine!

MECHANIZATION OF THE COAL INDUSTRY IN THE TWENTIETH CENTURY

By David A. Zegeer
Former United States Assistant Secretary of Labor

Recent public opinion polls conducted on behalf of the Kentucky Mining Institute have shown that the majority of the public still view coal mining as a pick-and-shovel, dirty, dangerous job, paying low wages in poor conditions for the health and safety of workers and the environment. What many people fail to realize is that mining is a high technology industry, and provides the basis for the very standard of living that everyone in this nation enjoys. More coal is mined by fewer people than ever before. The public opinion polls have revealed that this is considered a negative. At one time, mining was a very labor-intensive, back-breaking task in which miners used the simple hand tools of the day (pick, shovel, breast-auger or drill). Miners would dislodge coal and load it in cars that were, in many cases, pulled by animals out of the mine. This required large numbers of workers. Since many mines were located in isolated, rural areas, many companies, in order to develop new coal reserves, would often have to import workers, and construct new communities to house and care for the new miners. These communities were called "coal camps." Many companies prided themselves on the quality of life provided to their workers.

Mechanization of the coal industry has resulted in a work environment that has become progressively safer for the health and safety of the miners of today. The new machines utilized are all based on the basic principles used by the hand tools of old—cutting (pick), drilling (breast-auger), or loading (shovel). Machines today utilize electronic controllers, computers, remote controls, and in many cases, the workers are protected by steel and hydraulic-controlled supports in underground mines. Surface mines utilize giant earth-moving machines that sculpt the land to create and recreate natural habitats for vegetation and animal life.

Public opinion research also indicates that people still view coal companies as irresponsible—believing they leave devastated wastelands after their mining operations. Nothing could be further from the truth. The land itself is a valuable resource and land owners wish to return land to productive use once the coal is mined. Lands have been successfully returned to their original uses, creating forests, fish and wildlife habitats, and productive farmland. Lands have also been transformed into more complex uses, creating areas that can be utilized for commercial developments, housing, airports, golf courses, parks, recreation areas, and schools.

The history of the development of the coal industry has many very interesting stories. In many areas, whole communities were developed

Private collection of David Zegeer

Goat pulling coal car in Leslie County, Ky. 1920s.

Private collection of David Zegeer

Miners with cutting machine.

to provide places for workers to live and services for everyday life. Building materials, lumber, bricks and other materials were often developed from the surrounding area. Railroads were built to the area where coal mines and coal preparation plants were located. Many communities built in the early part of the century could be termed model towns—with homes, schools, churches, stores, hospitals, ice plants, power plants, water systems, recreational facilities, and farms.

Many miners in the early part of the century were recent immigrants. In order to train people, not only in mining coal, but also in first aid and safety, competitions between mines and companies were often held at local baseball parks. Mining towns, in those days, were famous for having their own baseball teams that would compete with each other, with many players going on to the professional leagues.

The mining of the coal, in the early part of the century, was simply a matter of digging "tunnels" through the coal seam and allowing the main roof to be supported by larger coal pillars. The coal was extracted as the coal seam was developed. On retreat, the larger pillars of coal were removed, causing the roof to fall. To mine coal in those days, six, seven, or eight "tunnel" entries were driven and connected by cross cuts or breaks for ventilation and haulage purposes.

The extraction of the coal meant dewatering the mine by pumping out water which would accumulate and hinder production. Pumping stations existed throughout the coalfield to pump this water out of the mine. In some cases, this water was used for drinking purposes.

Safety was always a major concern. In the early days, the detection of methane gas and low oxygen content was done by the use of canary birds. When the oxygen level decreased, the canary would pass out before miners, thus giving a warning. Later on, the flame safety lamp was developed to detect a low percentage of oxygen or an accumulation of combustible gas. Miners could judge by the brightness and intensity of

Private collection of David Zegeer

Coal recovery auger at Hazard Seam #4/Mine #24. October 19, 1961.

the flame how much oxygen was in the air.

The first coal seams were thicker and were mined by the pick-and-shovel methods. One of the typical working places for hand loading in those early days was where the rail was placed into the working face. A small mine car was pushed up and loaded by hand. Any rock partings that existed were carved out and separated in the workplace before the coal was sent outside to the tipple. Wood timbers were used for roof support. These consisted of vertical posts and what were called "cross-bars," which were used to support the rock overhead. These methods were the best that were available at that time, but still required much improvement.

Machinery was developed which lifted these heavy wooden cross-bars to the roof and allowed the legs to be placed beneath to support the roof. The use of this heavy equipment for roof support became more sophisticated year after year. Heavy hydraulic equipment was developed, diminishing the amount of manual labor required. The timber used in this method of roof support required that many trees be cut. Handling the timbers was a tremendous job, as was delivering them to the working place underground. At times, the weight of the roof, although supported by timbers and eventually, steel cross-bars, was heavy enough to bend even the steel cross bars. When the roof would fall and a passageway was necessary, removing the rock became a difficult and challenging job.

In the late 1940s, steel roof bolts, similar to toggle bolts that one uses around the house, replaced timbers. The steel bolts would penetrate the overlying rock in the roof. The laminated slate would be bound together by these roof bolts. Drilling these holes caused dust problems. Initially, respirators were used, and later, water was used while drilling to suppress the dust. In addition, collectors were installed to gather the dust through a hollow core in the drill steel to help minimize the dust emission.

Wedge type bolts were first used for roof bolting. They were installed by pushing them in against the top of the hole. A roof bolt plate was then placed on the bottom of the roof bolt, and a nut was used to tighten the plate against the roof.

Subsequently, the development of equipment for drilling holes and the use of roof bolts advanced further. The use of water in dust collection was a great improvement. Moreover, the use of this equipment accelerated to the point where today, massive pieces of equipment are used to install roof bolts, and two people can operate a dual roof drill in one working area.

Coal dust, when suspended in the air, can be explosive. To keep coal dust from exploding it is rendered harmless by the use of pulverized limestone rock which is called "rock dust." Miners apply the rock dust by hand to the coal, coating the entire surface area. Rock dusting equipment can also now spray large amounts of rock dust very quickly.

When preparing to shoot the coal and break it into small chunks for loading, one had to drill a hole to insert the explosives. The first method was the use of the breast-auger, a manual method of drilling similar to a brace and bit in a carpenter's workshop. Then, electric hand drills were used. Steps were taken to introduce hydraulic hand drills, operated from a cutting machine or roof drill. Still more improvements were needed. This led to the development of a mobile coal drill which was rubber-tire-mounted and maneuverable, and the operator could easily move it from one working face to the next. It decreased the amount of manual labor, which was another advantage of this mechanized process.

To avoid drilling a lot of holes and shooting the coal with more explosives, a second face was cut, or actually undercut, into the coal seam. In the early days, miners would lie on the bottom and cut with a pick. Cutting machines were de-

Private collection of David Zegeer

Miners with timbering machine, Mine #32. January 21, 1947.

veloped to speed up the process. The original ones were called "bottom cutting machines," and they would make a six to eight inch cut, which would allow one to blast the coal more effectively. Cutting machines look like giant chainsaws. Later, rubber-tire-mounted cutting machines were developed which could maneuver about more easily in a mine. One could also rotate the cutting bar to cut a vertical slice in a coal seam in addition to a horizontal cut. These rubber tire cutting machines were quite an improvement over what was used initially, but they still presented the problem of coal dust. To minimize the coal dust, water was piped directly into the working place and through the cutting machine.

As the coal drilling was completed, the person who shot the coal, a shot fireman, would insert the explosives. He would insert the explosive material in the holes, then, us-

Private collection of David Zegeer

Miner with portable drill in transit, 1950s.

Private collection of David Zegeer

Miner with portable drill in operation, 1950s.

ing a blasting unit, detonate this explosive, shooting the coal down, and making it ready to be loaded with a loading machine.

In those early days, in low coal, one of the means of loading the coal was called a "duck bill," which was just a conveyor that would slide under the coal, shake it back into a pan and down to a conveyor line. This method was further developed into what is called a continuous haulage system. Instead of using shuttle cars, a continuous haulage system would shuttle the coal from the loading machine back to either a mine car or a belt conveyor. These continuous conveyors are still in use today, particularly, in thin coal seams.

The loading machine, which was developed in the 1930s and 1940s, loaded the coal in the shuttle cars and transported it from the loading machine to a point where the coal was hauled out of the mine. Most shuttle cars had four wheels that would support them. Some were developed with six wheels. Today, shuttle cars have been developed to haul as much as 15 tons of coal on each trip to a dumping point. The shuttle cars, in conventional mining or even with continuous mining, are still one of the essential pieces of equipment for hauling coal.

The coal is then dumped from a shuttle car onto a belt feeder, a machine that feeds the coal onto a belt conveyor. This stage of the process provides a uniform method of feeding the belt. In later developments, a crusher was incorporated on the belt feeder to break large lumps of coal and to prevent them from spilling off the belt conveyor. These feeder-breakers are in use today and are usually located where shuttle cars are dumping coal onto a belt conveyor. This is another example of how efficient the development of mining equipment has become in the haulage of coal.

Following conventional mining techniques in the late 1940s (cutting, drilling, shooting, and loading), the development of continuous miners evolved to incorporate all these steps into one process of just ripping coal from the working face, loading it into a shuttle car, and hauling it back to the dumping point. Continuous miners have been built by manufacturing companies such as EIMCO, Joy, Jeffery, Fairchild, and Lee Norse. They incorporated many steps in the mining process to enable companies to remain competitive by achieving a higher percent of recovery while enhancing the safety of the miner. The conventional mining system is rarely used in this country today.

Another major innovation in continuous mining was the introduction of longwall mining. This entailed either a large shearing machine or a plow that literally rips the coal from the working face. A hydraulic shield roof support was designed for the machine which would protect the workers and equipment as the coal was being extracted from the working face and loaded onto a conveyor which transported the coal outside the mine. The working face can be several hundred feet across. These hydraulic supports and longwall shearers are very expensive components. To buy units of equipment such as these, the cost could escalate to tens of millions of dollars. A large reserve of coal is required to justify the purchase and installation of these sophisticated machines. Blocks of coal several thousand feet long can be mined, with miners standing below steel canopies.

As the shearing machine advances, the support is the shield or the steel chock located above the operator and equipment. The shield advances with the equipment and provides protection for the miners from roof falls. In summary, longwall mining advances and rips the coal from the face. A shield protects the workers and the equipment from roof falls during the coal extraction process. The coal is hauled on a belt conveyor where it is loaded into mine cars, transported to the outside, and finally, conveyed to the preparation plant.

Mechanization of the Transportation of Coal

The modern conveyor today is quite an advancement from the primitive methods of the early days when coal was hauled by oxen teams, mules, donkeys, ponies, dogs, or even "billy goats." The small rail mine cars would hold about one to two tons of coal, and all of it was hand-loaded. Later, electric locomotives were utilized to pull these mine cars. More progress was made with the introduction of larger mine cars that hauled ten tons as opposed to the two-ton mine car, thereby justifying the expenditure of installing a better haulage system. These cars were loaded directly from shuttle cars or from a belt conveyor at automatic loading points. Subsequent developments introduced the 25 and 30 ton mine cars and the coal was loaded automatically into these cars. To accommodate the use of these large cars, more space was required in the mines. It took a lot of rock excavation and tunneling to accomplish this.

Private collection of David Zegeer

Miners using pneumatic wrench on roofbolt, 1947.

Personnel Transport

Hauling coal was not the only problem in transportation. Getting personnel to the working face posed another problem. Man-trip cars were used

Private collection of David Zegeer

Miners in mantrip, 1950s.

to haul workers, crew-by-crew, to the working places in the mine. In thin coal, even personnel mobility was a problem within the mine. Little units called trikes, mine buggies, or mine mouses were used for individuals traveling around in low coal.

Surface Mining

Underground mining can get just so far to the outcrop or surface. Surface mining utilizes the same types of equipment and processes that are used in highway construction or any other earth-moving project. Some of the earliest surface mining occurred in the first part of the 20th century when steam shovels used to build the Panama Canal were brought back to this country. In the mid- and late 1940s, surface mine equipment included shovels, bulldozers, rock drills, and other equipment to mine coal that was left by underground miners. Surface mining was and still is quite controversial. Coal that could not be surface mined or mined underground was augered. Augering is similar to the use of the brace and bit found in a carpenter's workshop. Augering would bring the coal to the outside where it was loaded onto trucks and taken to a preparation plant or railroad loading point. Today, a technique that uses underground continuous miners to bore holes from the surface is known as highwall or thin seam mining.

One of the great concerns with surface mining was the reclamation of the terrain. Surface mining and reclamation has improved through the years but remains an emotional issue. Different methods have been used in reclamation and the problems presented brought about further advances in reclamation of mined land, such as providing lakes and ponds, as well as level acreage for community development in the mountains. There are numerous facilities, such as schools, hospitals, airports, prisons, golf courses, and new housing developments built on reclaimed surface mined land.

Work Force

A common practice in the early part of the 20th century was to include children as part of the work force. In later years, the work force included not only laboring people, but also engineers, well-trained technicians, and other capable persons who were independent and endowed with a great deal of common sense and formal education.

Preparation Plants

Once the coal was mined, quality improvements were made in the use of preparation plants. Preparation plants were developed to provide a simple way of washing the coal and separating it from rock with the purpose of providing a good quality of coal for the market. Early plants, often called tipples for the mine cars that were unloaded by tipping them over, used hand labor to pick rock from the coal.

Preparation plants of today use sophisticated chemical separation processes controlled by computers with nuclear sensing devices to weigh and analyze the product in process. Large quantities of rock and coal fines are left over which are placed in engineered fills and impoundments.

Virginia Tech Corporate Research Center, Blacksburg, Virginia

1960's Long-Airdox loader with piggyback.

Virginia Tech Corporate Research Center, Blacksburg, Virginia

1990 Long-Airdox Model 728 continuous miner.

In summary, mechanization of the coal industry since the early 1900s has contributed to the rapid industrialization of this nation. During each phase of development, problems were addressed by more innovative mechanization and creative miners who encountered these problems. The use of technological advances in mining has fueled our nation as it developed throughout the twentieth century and into the next millennium.

(Adapted from a non-technical presentation given with a slide show to a mining group in Lexington, Kentucky and Denver, Colorado, by David A. Zegeer, P.E.)

Komatsu America International Company

Komatsu D375A Bulldozer at surface mine.

Brandeis Corporate Office

Ingersoll-Rand Drill, September 29, 1995.
Brandeis Heavy Equipment, Louisville, Ky.

ADVANCES IN PROTECTIVE EQUIPMENT

By Elmer Howard
Assistant General Manager—East Ky. Div., Alliance Coal, LLC

Many dramatic changes have been made to the miner's personal equipment during the 20th century in Kentucky's coal mining industry. At the beginning of the century, underground coal miners wore no special protective equipment. A canvas cap with a leather bill was the only article of clothing that differed from the clothing that farmers or factory workers wore to work each day. This cap provided little or no protection to the miner. Its primary purpose was to support an open-flame light used to illuminate an underground passageway that would otherwise have been void of light.

The canvas cap the underground miner wore at the beginning of this century gave little protection. At best, this cap would protect the miner from small cuts or abrasions. It provided no protection from portions of the mine's roof that gave way from time to time. Clearly, more protection was needed for the miner's head.

While providing necessary light, the open flame lamp created a hazard with potentially fatal results for the underground miner. Methane gas is found in all underground coal seams. This gas is trapped inside the coal. When coal is mined, the quantity of gas that is liberated into the mine atmosphere varies greatly within each seam. Open-flame lights are a source of ignition for methane when the gas is allowed to accumulate in an underground mine.

If a miner was able to survive the initial explosion of methane gas, he had no guarantee of safe passage to the surface. Carbon monoxide is produced when methane ignites. This becomes a deadly atmosphere after the fire has subsided. Early miners had no protection from a noxious mine atmosphere.

In the early days of mining, miners had no special protection for their feet. Leather, lace up "brogans" were worn by most underground coal miners for over one-third of the 20th century. Falling materials presented potential for foot injury on a daily basis for underground coal miners.

During the early part of the 20th Century, underground-mining-related injuries and fatalities increased as the demand for coal increased. Coal fired the furnaces that produced steel which, in turn, was used to manufacture consumer goods and military materials. As the demand for coal increased, it became the primary fuel used to produce steam and drive turbines that produced electricity. As the United States grew in population, so did the demand for an economical fuel source and coal became the preferred choice.

As improvements in mass communication developed, the public became more aware of the

Private collection of David Zegeer

Coal miner of the 1950s.

number of injuries and deaths in the underground mining industry. Soon afterwards, the public demanded a safer workplace for the miners. Purchasing coal from coal companies that had a high frequency rate of mining related deaths and injuries became socially unacceptable.

Finding solutions to the problem of injuries and death produced a prime example of technology being invented from necessity. In order to solve the problem, it was necessary to determine the causes of injury and death in mines. Accidents were investigated to determine the causes and information was compiled in an effort to determine what safety tools and practices were needed to reduce pain, suffering, and death.

The personal equipment used by today's underground coal miner has very little resemblance to their turn-of-the-century predecessors. Contemporary miners enter an underground mine equipped with a hard hat, battery powered light, hard toe shoes, heavy leather work gloves, a rescue breathing device, and gas detection devices.

As the demand for coal increased in this nation, underground coal miners went farther and farther into the earth's crust to retrieve this valuable source of energy. As a rule, the deeper into the earth miners penetrated, the more likely they would be faced with a highly explosive methane gas. This gas is colorless, odorless, lighter than air, and can be ignited by explosives, open flames, or sparks. Because of the physical properties of methane, early miners were presented quite a challenge in detecting this potentially deadly gas. Early on, caged canaries were brought into the underground mine. In a noxious atmosphere, canaries would expire long before men. As long as the canaries lived, miners believed it was safe to enter working areas of the mine. Some mining companies practiced burning out the gas before working in an area. This method provided some measure of safety for the miners.

One of the most effective means to detect methane was the invention of the flame safety lamp. An open flame protected by wire mesh and glass was used by underground coal miners to determine the amount of methane present in a mine's atmosphere. The construction of the flame safety lamp allowed a flame to burn inside a device that would not normally ignite methane in the mine atmosphere. Trained users of this lamp were able to determine the presence and percentage of methane in a mine by observing the lengthening and color of the flame inside the lamp. This type of detection device is still in use today, although electronic sensors are quickly replacing this technology.

Martin County Coal Corporation

Machine operator Benny Salyers at White Cabin Mine. Martin County Coal Corporation, Inez, Ky.

As the study of electricity and electronics progressed during the 20th century, other means became available for the detection of methane. Today, a hand-held electronic instrument is used to detect methane in underground mines. These methane detection devices have also been installed on mining machinery. These detection devices, when working properly, will not allow a machine to operate when methane levels rise above a predetermined level. Modern electronic methane detection instruments are able to detect methane in minute quantities, well below explosive range.

Light is of the utmost importance to underground coal miners. The absence of light renders the underground coal miner helpless.

Martin County Coal Corporation

Elmer Howard.

In the beginning of underground mining, light was produced by an open flame. But an open flame presented the distinct possibility of igniting methane gas. Many attempts have been made over the years to protect the open flame from the surrounding atmosphere. With advances in the study and application of electricity, battery powered lights eventually emerged and can be safely used in underground mining.

Earlier in the 20th Century, the failure of safety devices and a subsequent methane explosion meant chances were not good that a miner would be able to retreat to a safe atmosphere. When methane ignites, carbon monoxide is produced. Carbon monoxide is a colorless, odorless, and tasteless gas that has deadly effects on the human body—even in very small quantities. Strangely, the human body prefers carbon monoxide three hundred times more than oxygen. Without life-sustaining oxygen, the human body will perish in a short period of time. Underground coal miners had a need for a device that protected them from carbon monoxide. Necessity, again, prompted technology.

First generation rescue breathing units for underground coal miners filtered out the carbon monoxide from the mine atmosphere. These devices have saved many lives in the coal mining industry. One shortcoming of these devices is that no matter how much carbon monoxide is filtered out if the proper amount of oxygen is not present a miner cannot continue to live. Today's rescue breathing units are fully self-contained. Oxygen is chemically produced and is independent of the mine atmosphere. Utilization of the principles of chemistry has played a role in developing lifesaving devices.

Head injuries were fairly common in the early years of coal mining. There was clearly a need to protect the miner's head. The miner's hard

James B. Goode, Photographer

Coal miner in bath house at Miller's Creek Mine April 1999, Lodestar Energy, Pikeville, Ky.

James B. Goode, Photographer

Mine foreman outside bath house at Miller's Creek Mine, April 1999. Lodestar Energy, Pikeville, Ky.

hat of today can withstand heavy impact forces without breaking or puncturing. This quality comes from the evolution of plastics. Today's non-metallic compounds rival both wood and metal in strength and endurance. These compounds have made it possible to construct a protective helmet for coal miners that is similar in strength to that of professional football players. One of the key elements in the added protection of today's hard hats is the suspension contained in the underside of the hat. The science of physics has taught designers to configure the suspension of the helmet to spread forces over the surface of the head rather than allow impact forces to be concentrated on one area. A combination of the study of physics (applied forces) and chemistry (compounds) has resulted in an increased measure of safety for today's coal miner.

Improvements in safety shoes have also evolved from necessity. Falling material from a mine's roof or ribs, movement of mechanical equipment, constant use of moving hand tools, and the potential injury from work animals all exposed coal miners to foot injuries. Foot injuries were the most common injuries among coal miners early in the 20th century. In the mid-part of the century, research revealed that adding a steel plate near the toe area of shoes prevented many devastating injuries to coal miners' feet. With new technologies being advanced during the latter part of the industrial revolution, it became relatively easy to form steel into a shape that would fit near the toe of work boots. These "steel toes" were originally placed on the outer part of

James B. Goode, Photographer

Bath house hangers at Miller's Creek Mine, April 1999, Lodestar Energy, Pikeville, Ky.

the work boot. Eventually, the steel plates were layered between two pieces of leather. This small piece of formed metal eliminated countless numbers of foot injuries.

More progressive companies have employed the use of metatarsal footwear in the workplace. Metatarsal footwear incorporates the use of the steel toe and adds additional protection farther along the length of the foot. Again, physics played a very important role in the development of metatarsal footwear. By spreading forces over a larger area, less force is applied at specific points on the foot, thus less damage is incurred to the wearer of the shoes.

This same concept is now being applied to gloves worn by miners. Padding is added to the back of gloves and extends from the wrist to the tip of the fingers. By allowing forces to be spread over a larger area, the likelihood of injury to the hand and fingers is decreased. Several underground mines in Kentucky employ the use of these metacarpal gloves.

During the mining process, coal must be broken to extract it from the seam. When coal is broken, small particles of coal dust are released. Today's modern mining machinery filters out much of this dust. Filtration of the air coupled with currents of air sweeping out the coal dust and directing it away from miners provide for a healthier atmosphere in which to work. In addition, ventilating current and filtration respirators are made available to the miners. The latest technology in respirators has produced very highly efficient devices. The most efficient of today's

James B. Goode, Photographer

Coal miner underground at Miller's Creek Mine showing protective miner's cap, April 1999, Lodestar Energy, Pikeville, Ky.

James B. Goode, Photographer

Coal miner underground at Arch of Kentucky's #37 Mine showing fan-vented protective cap for longwall section, 1995, Arch of Kentucky, Cumberland, Ky.

respirators have the capability of filtering out 99.97% of dust particles from the atmosphere.

A new design of hard hat with an air filtration system is in use in some underground mining applications. This system uses a battery powered fan that pulls air into the back of the hard hat, filters the air, and forces it out vents near the forehead area of the helmet. A clear shield is attached to the helmet and the air flows downward on the inside of the shield. With this technology, the miner is supplied with a constant stream of filtered air which contains little or no dust.

The study of physics, chemistry, and electricity has led to improvements in safety for coal miners on the job. Without the creativity and ingenuity of inventors, coal miners would still be wearing canvas caps, "brogans," and carrying open flame lights. Highly impact resistant mining helmets, steel toe shoes with metatarsal guards, metacarpal gloves, safe battery powered lighting, self-contained self-rescuers, respiratory protection, and highly sophisticated gas testing instruments all have contributed to the reduction of pain, suffering, and death in today's mining industry.

UNDER THE GROUND

By Steve Gardner
President
Mining Consulting Services, Inc.

Underground mining is used when the coal is buried too deeply in the ground to make surface mining feasible. There are three kinds of underground mines: shaft mines, drift mines, and slope mines. The type of mine that is developed depends upon the terrain and other factors.

Slope mines are used to reach coal beds in areas where the seam lies near the surface but is too far underground for surface mining. A tunnel is slanted downward through the ground to the coal. The miners normally travel in and out of the mine by vehicles called portal buses or mantrip cars. Coal is taken out of the mine by rail in mine cars or by conveyor belts.

Shaft mines reach coal beds that lie far below the earth's surface. A shaft, or hole, is excavated straight down to the coal seam. The miners then develop horizontal entries (tunnels) through the seam of coal so that the coal is mined through tunnels spreading out from the shaft. Elevators, similar to those used in tall buildings, lower or raise miners and equipment between the mine entrance at the surface and the level of the coal. Other elevators or mine hoists lift the coal out of the mine.

Separate shafts are normally excavated to provide ventilation for the underground mining operation. Typically, many shaft mines in the United States range from 200 to 1,000 feet, but some mines are as much as 1,500 feet deep. In Europe, there are shafts nearly 4,000 feet deep.

Drift mines are used to develop coal beds exposed in hillsides. The mine entrance is located horizontal to the coal seam. Miners tunnel farther and farther back into the hill, removing coal as they go. Electric locomotives or rubber tire vehicles carry men and equipment in and out of the mine. "Mantrips" carry people into the mine and pull supply cars. Small utility vehicles called scoops carry supplies and clean up loose coal in the mines. Some mines use diesel equipment like the HumVee used by the Army. Sometimes conveyor belts are used to transport the coal. This type of mine is generally more economical to open than a shaft mine because excavation through rock is not normally required.

Sometimes an individual mine will have all three types of openings—drift, slope, and shaft. For example, the coal haulage might come to the outside through a drift opening when the preparation plant is nearby. As the mine develops and expands, additional openings may become necessary for ventilation and to shorten travel time for personnel and equipment. A slope opening might then be used for the portal where the cover is not too great, and a shaft, or shafts, might be used for ventilation.

Complex Underground Machines

Coal seams in Kentucky generally lie in flat layers under the ground. In the mountains of eastern Kentucky, many coal seams occur from the top of the mountain all the way down into and below the valleys. In western Kentucky, there are numerous seams of coal that underlie the surface of the land. Dozens of seams are found along a length of several hundred feet. Coal seams are known by localized names, such as the Harlan, the Elkhorn, the Blue Gem, the Number 9, the Hazard, the Whitesburg, etc. Coal seams have been mined underground as thin as about 24 inches up to several feet thick. A typical coal seam being mined today can range from the low 30-inch range to 60 inches or more.

To begin mining, a portal for the mine

Consol Coal Group, Pittsburgh, Pennsylvania

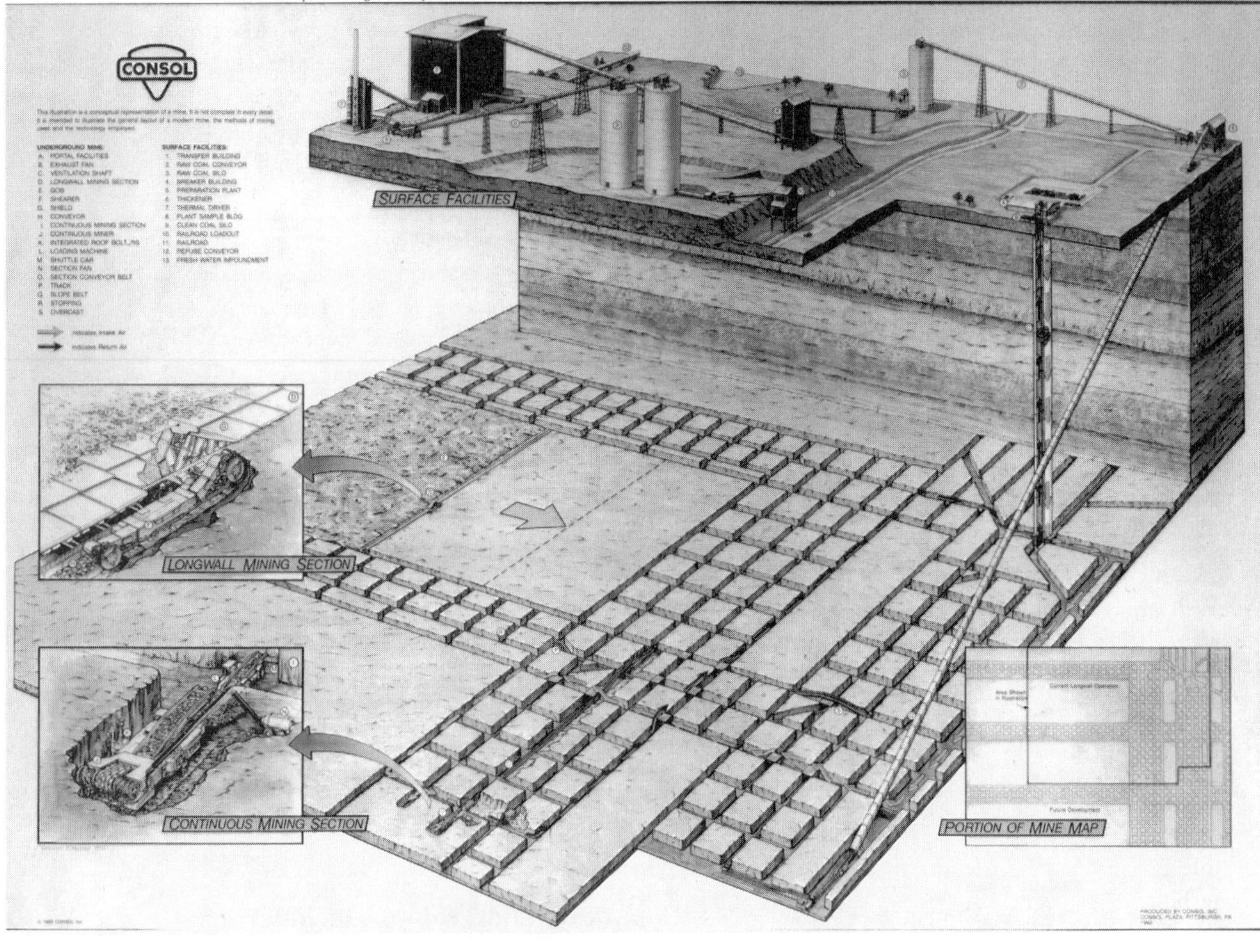

Diagram showing cross section of underground mine.

face-up is excavated. In areas where the seam is above drainage (that is above streams in the area), the point where the coal seam hits the surface is known as the outcrop. Coal is mined by excavating tunnels or entries straight ahead with intercepting cross-cuts going to the sides and leaving blocks of coal or pillars that help hold up the roof. Bolts are used to hold up the rock above the entries or cross-cuts where coal has been removed. These roof bolts are placed in holes drilled vertically and act much as nails or screws holding wood together. Some bolts are mechanical and some use resin or glue. This process of roof-bolting pulls the layers of rock together, much like pieces of plywood or laminated wood products are held together. Air is forced into the mines with large fans to provide fresh air for miners to breathe and sweep away the dust created by the mining process.

The machines used to mine the coal utilize a large, rotating cutter that breaks up the coal and loads it on a conveyor where it is transferred to shuttle cars or directly to a conveyor belt that brings the coal to the surface of the ground. Many of these machines, called continuous miners, are controlled by the miners with radio remote controls.

Consol Coat Group, Pittsburgh, Pennsylvania

Miners operating longwall miner.

Roof-bolts are placed in the roof of the mine with a machine called a roof-bolter, after a hole is drilled upward and into the roof. A long bolt of at least thirty inches and sometimes several feet is placed in the hole and tightened.

In a large mine, a process known as longwall mining results in large volumes of coal being mined. In this process, the continuous miner lays out long, wide panels of coal by the room-and-pillar method. Large hydraulic jacks which can support thousands of tons are placed in a line sometimes as long as a thousand feet. A shearer which has a rotating cutter head moves back and forth along the face of the coal, allowing it to drop onto a conveyor which transports the coal to a conveyor belt leading out of the mine. This method of mining can result in thousands of tons of coal per day. As the coal is mined, the large hydraulic jacks move forward. Overlying rock

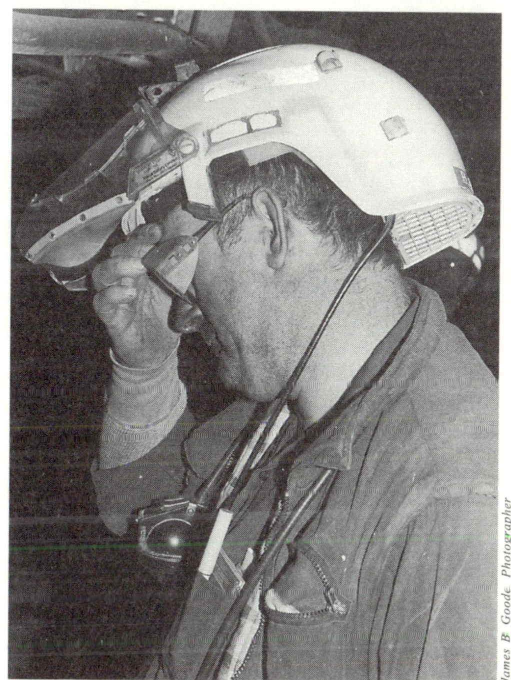

James B. Goode Photographer

Coal miner underground at Arch of Kentucky's #37 mine showing fan-vented protective cap for longwall section, 1995, Arch of Kentucky, Cumberland, Ky.

Martin County Coal Corporation

Continuous mining machine.
Martin County Coal Corporation, Inez, Ky.

then falls behind the jacks with the workers safely under the large steel canopies. In some areas, where the rock falls, the rock layers above the mine can form cracks and the surface of the land can be lowered or subsided. This is planned for in advance; however, sometimes cracks can develop that require work to be done on the surface to fill them.

Martin County Coal Corporation

Miner operating roof bolting machine. David Branham (left) and Howard L. Litton (right). Martin County Coal Corporation, Inez, Ky.

Martin County Coal Corporation

Rodney W. Tackett operating roof bolting machine at Pilgrim Mine.
Martin County Coal Corporation, Inez, Ky.

Martin County Coal Corporation

Miner operating shuttle car.
Martin County Coal Corporation, Inez, Ky.

Martin County Coal Corporation

Miner operating shuttle car.
Martin County Coal Corporation, Inez, Ky.

ON THE SURFACE

By Steve Gardner
President
Mining Consulting Services, Inc.

Surface mining moves the overlying dirt and rocks from a coal seam. In mountainous areas, the coal seams can be removed by contour mining which follows the outcrop of the coal seam around the hill, creating a bench from which the coal is removed. The bench is then backfilled to the "Approximate Original Contour" or AOC. In some areas, an auger, a large screw-like machine, is used to bore into the sides of the hills to extract additional coal. Highwall miners, using a machine similar to the continuous miner in underground mines, cut out long rectangular holes into the side of the hill. On the tops of mountains, the overlying earth and rock known as overburden can be removed sometimes as deep as two or three hundred feet, and several seams can be taken at one time.

Clearly, coal cannot be surface mined without changing the appearance of the land. However, all sites are mined with reclamation in mind. It must be remembered, however, that there is no such thing as "instant" reclamation. Successful reclamation of mined land takes time for seeds to germinate, for root systems to develop, and for vegetation to take hold and thrive.

Every phase of surface mining, from the initial cut to final grading and seeding, is planned and controlled. For example, the width of a cut, or bench, that can be made in a hillside depends on the degree of steepness of the slope, according to mining regulations or engineering criteria. This, in turn, limits the amount of earth that can be displaced. Even before the first bulldozer displaces a single cubic yard of earth, reclamation plans must be submitted to appropriate government agencies for approval.

James B. Goode, Photographer

David Duff, owner Pine Branch Coal Sales at Combs' Branch, Red Rock #10 Surface Mine, April 29, 1995, Pine Branch Coal Sales, Chavies, Ky.

Methods of Surface Mining

There are four methods of surface mining: area mining, contour mining, open-pit mining, and auger or highwall mining.

Surface mining is the process of removing the layers of earth and rock that cover the coal when the coal bed is close to the ground's surface. When the coal is exposed, it is broken up and loaded into trucks. Then it is transported to a preparation plant or loaded into a railroad car for direct delivery to a customer. In most cases, surface mining is a faster and more efficient way to mine coal near the earth's surface.

Area mining is usually practiced on relatively flat terrain where giant draglines and shovels strip away the earth and rock above the coal. This type of mining is conducted over large areas of land. In mountainous terrain, a modified method of area mining, known as mountaintop mining or mountaintop removal, uses the same techniques. To begin, pre-stripping operations with smaller equipment create a large flat area for the draglines or shovels to operate.

After the coal is removed, the tops of the mountains are reclaimed to flat or rolling contours, creating areas of usable land in the mountains. Excess rock from the top of the mountain is placed in structures known as hollow fills, which are very similar to the road fills along the sides of highways.

In western Kentucky, where the ground is flatter and more hilly than mountainous, area mining is used. Long open pits are excavated for the coal to be removed. Rock is excavated from mining and placed in the open pit behind the machine.

Contour mining is commonly practiced in mountainous terrain where the seams of coal outcrop on mountain slopes. In contour mining, the earth and rock are removed from the mountainside in order to expose and remove the coal.

Open-pit mining is most often used where the coal beds are extremely thick or where they are steeply sloped. Huge power shovels and trucks remove overburden in levels, called benches.

Then the trucks haul all the material out of the pit. Once the coal seam is finally exposed, shovels, trucks, and sometimes draglines are used to remove the coal.

Auger or Highwall mining is similar to contour mining in that it is practiced in mountainous terrain where the seams of coal outcrop on the slopes. Dirt and rock are removed to create a bench and expose the coal. The coal is then removed using giant augers to bore long holes into the coal seam. Augers are boring devices similar to drill bits. Highwall mining utilizes a modified continuous mining device which operates by remote control and removes the coal by a rotating drum of smaller solid steel bits.

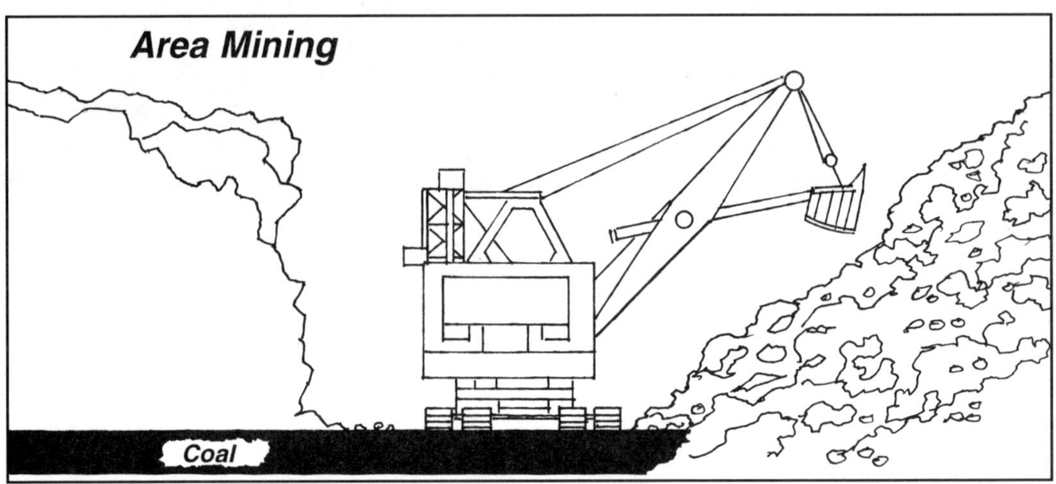

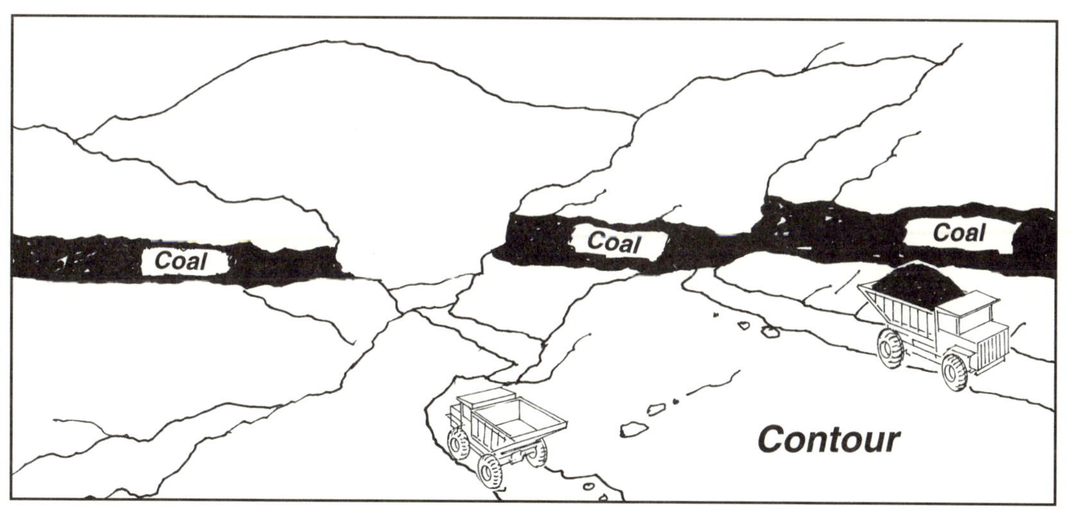
Coal
Coal
Coal
Contour

Coal
Coal
Open pit Mining

Coal
Coal
Augar or Highwall Mining

Simple and Complex Machinery for Mining on the Surface

Various pieces of simple and complex machinery are used in the process of mining coal on the surface. These include devices which drill, rip, scrape, scoop, and smooth. Bulldozers are used to push the earth and rock away from the coal seams and then to level the backfill after mining is done. Although many in the mining industry refer to the material that is left after mining is completed as "waste," it is actually soil and rock used to backfill areas where the coal has been removed. Some bulldozers used are as tall as a one-story building. Track mounted drills are used to make the holes that are filled with explosives that break up the rock so that it can be moved. End-loaders have buckets that pick up the earth, rock and coal to be loaded into trucks for transport. Backhoes aid with building ditches, ponds, and roads. Drainage from the mine sites is diverted into sediment ponds, where the majority of sediment or dirt is allowed to settle to the bottom of the impoundment. Road scrapers smooth the road surface and assist with ditching for proper water drainage.

Large haulage trucks haul overburden from above the coal to the fill areas or backfills. The large trucks can carry almost 300 tons of rock. Global Position System (GPS) units that receive signals from satellites are placed on some trucks so the location of a truck is known at all times. Remote control equipment is also under development and being used at some mines. Training devices similar to aircraft simulators are used at some operations to train the equipment operators.

Many innovations are being employed in mechanics and electronics. Additionally, computer technology and attendant software have become increasingly important in the deployment of surface mining equipment.

James B. Goode, Photographer

A Kentucky legislative task force pose in front of Komatsu D575A Super Dozer at Ivy Creek Surface Mine, April 1999, Lodestar Energy, Pikeville, Ky.

Innovative Mining & Equipment, Gillette, Wyoming

225,000 lb. 450 H.P. Turbo Charged M-880 Multi-Wheel Dozer, February 1999

James B. Goode, Photographer

Komatsu D575A Super Dozer at work at Ivy Creek Surface Mine, April 1999, Lodestar Energy, Pikeville, Ky.

Rudd Equipment Company/Roberts Communications, Inc., Pittsburgh, Pennsylvania

Tamrock Blasthole Drill.

Hitachi Construction Machinery (America) Corporation, Houston, Texas

Super EX5500 Hitachi Excavator, 1999.

Komatsu America International

Komatsu Demag H455S 2,250 H.P. Hydraulic Mining Shovel, September 11, 1996.

Martin County Coal Corporation

Bucyrus Erie Shovel,
Martin County Coal Corporation, Inez, Ky.

Bucyrus Erie Shovel
Martin County Coal Corporation,
Inez, Ky.

Martin County Coal Corporation

Charolais Coal Corporation

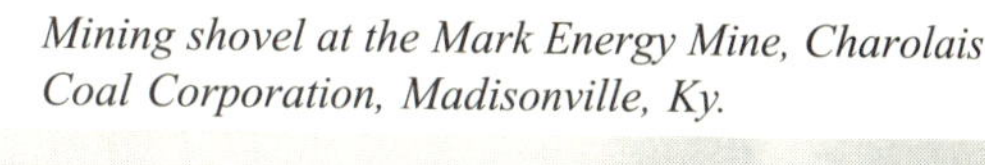

Mining shovel at the Mark Energy Mine, Charolais Coal Corporation, Madisonville, Ky.

630E Dresser Haulpak Truck, Martin County Coal Corporation, Inez, Ky.

Martin County Coal Corporation

KEY CONSIDERATIONS IN COAL TRANSPORTATION: PREPARING FOR THE TWENTY-FIRST CENTURY

By Joseph E. Lema, Vice President
Manufacturers and Services Division
National Mining Association

Railroad Transportation of Coal

For many years, coal has been important to railroads. Although coal is no longer used to produce steam for their locomotives, today it represents the leading product carried by railroads. Most of this coal is carried to power plants that generate electricity for industrial, commercial, and residential consumers. Coal is the low-cost fuel of choice used by utilities to generate 57% of the nation's electricity. Coal amounts to approximately 45% of the freight tonnage carried by the leading railroad companies in the United States. Because nine out of every ten tons of coal produced in the United States are consumed for generation of electricity, it is clear that railroads play an important role in the distribution of coal from mines and processing facilities to power generators.

In 1997, the major U.S. railroads (CSX Transportation, Norfolk Southern Corporation, Burlington Northern Santa Fe Railroad, and Union Pacific Railroad) distributed 153 million, 116 million, 193 million, and 160 million tons of coal respectively. These companies accounted for the movement of approximately 96% of the annual coal freight tonnage in 1999.

Power generating facilities consume 90% of the coal produced. Less than 10% is consumed in other markets. These include non-

CSX Transportation, Jacksonville, Florida

CSX coal car being tipped and unloaded.

utility industrial facilities or exports to foreign consumers such as Canada and Mexico. The percentage of coal hauled by eastern and western railroads is evenly divided. This demonstrates a dramatic rise in the production of western coal in the two decades since the 1970s. Eastern coal production has remained much thc samc during the corresponding period.

Much of the increase in coal-fueled power generation has been supplied by coal from new and expanded coal mining operations in western states such as Wyoming, Utah, and Colorado which are often long distances from population centers where electricity is generated. This requires even more shipments by railroads. Demand for electricity rises 1.5 to 3.0 percent every year across the U.S. At the advent of the 21st Century, a growing population and an expanding economy have created more demand for coal-fueled electrical generation. As a result, rail transportation in the 21st century plays an increasing role in the coal industry.

Major consolidation within the rail industry has occurred in the past two decades. Government deregulation has helped make these mergers easier. Most of these have come from connecting rail lines where one ended and another began.

Virginia Tech Corporate Research Center, Blacksburg, VA

Long-Airdox 1990s full dimension continuous haulage chain.

Continental Conveyor & Equipment Company, Winfield, Alabama

Underground coal conveyor of the 1970s.

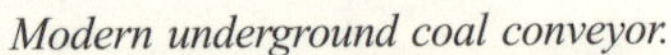

Modern underground coal conveyor.

Continental Conveyor & Equipment Company, Winfield, Alabama

Railroad Mergers To Form Class I Railroads

*CSX Transportation (CSXT) — Chesapeake & Ohio, Baltimore & Ohio, Western Maryland, Seaboard Coast Line, Louisville & Nashville, Clinchfield

*Norfolk Southern Corp. (NSC) — Norfolk & Western Railway, Southern Railway

Burlington Northern Santa Fe Railroad (BNSF) — Burlington Northern, Atchison, Topeka, & Santa Fe Railroad

Union Pacific Railroad (UP) — Union Pacific, Missouri Pacific, Southern Pacific, Chicago & Northwestern, St. Louis Southwestern, Denver & RioGrande Western

**Canadian National (CN) — Canadian National, Illinois Central
**Canadian Pacific (CP)
**Kansas City Southern (KCS)

* CSXT and NSC each also acquired segments of Conrail in 1999.

** CN, CP, and KCS handle some coal traffic, however much less than CSXT, NSC, BNSF, or UP

Some short-line regional railroads also carry some coal traffic to connect with Class I railroads on feeder tracks.

The result of these mergers and consolidations is that the number of railroads has declined sharply and left much of the long-haul coal business in the hands of four major companies—CSXT and NSC in the east and BNSF and UP in the west.

Reform in the regulation of railroads by the government has benefited the industry. They are returning more money to their investors and are charging less to move the coal. Several parts of the Staggers Act of 1980 have led to this improvement. Railroad companies have been allowed to establish contracts with the coal companies which allows for individual negotiation. This enables the railroad company to better manage its operation by tracking their crews and equipment and using them to their best advantage. Other advantages include 1) The formation of more single line service which can more efficiently move coal for long distances. 2) Increased productivity as a result of reducing the number of employees 3) Cutting costs by retiring locomotives which had become outmoded and adding newer, more efficient ones.

Cyprus Mountain Coals / Tim Cox PhotoGraphics, Norton, Virginia

Shovel Loading Dresser Haulpak 630E Truck.

What does the 21st century hold for railroads and shippers?

Railroad companies have demonstrated that they are capable of handling more coal. Presently,

CSX Transportation, Jacksonville, Florida

Coal unit train being loaded at tipple.

they are hauling 162 million more tons than in 1988. In 1997 there were 482 billion ton-miles of coal transportation service. This represents a 164 billion ton-mile increase since 1988.

Railroad companies will invest more in modern, AC-type locomotives and increase the number of available crews to meet higher demand. Improvements will continue to be made in the railroad infrastructure and traffic control. Bulk products such as coal and grain and non-bulk container traffic will increase along with the rising national economy.

Communications systems will be improved to assure reliable movement of coal from mines and processing plants to destinations where coal is used. Train dispatchers, train crews, mine supervisors, and power company officials will improve their day-to-day business communications. At the beginning of the 21st Century, all the major players in the coal production and transportation industry will realize that an efficient energy delivery system is becoming more and more vital to the economy and security of the nation. As a result, coal will be ready for shipment when the coal train arrives; the coal will be hauled and transferred with speed and safety; coal stockpiles at power plants will be maintained at satisfactory levels during times when the peaks and ebbs of power usage are most likely.

In the United States, the current known coal reserves will allow an estimated 270 years of mining at 1.2 billion tons of coal per year. That is over 324 billion tons of coal! This means that the efficient transportation of coal will remain critical for many years to come.

Barge transportation of coal

The United States has an 11,000-mile-system of connected inland and intracoastal waterways which can be navigated. These have long served to move coal from terminals to eastern and midwest utility and non-utility power generating plants. Some of this coal is transported to the Gulf of Mexico and several points on the lower Mississippi River. Some railroad systems carry coal from the west to the Mississippi River in Illinois and Missouri. Additionally, coal is brought down the Ohio River to where it joins the Mississippi.

Barge transportation has provided an effective alternative for many coal producers. Many of the power plants are located on navigable rivers which are easily accessible by barge. The ports located on the gulf from which coal can be exported are also easily accessible

Cyprus Mountain Coals

Belt loading 18 wheel transport truck.

from the rivers, intracoastal waterways, railroads, and other avenues of transport from inland United States. There are many modern coal terminals in place along the nation's waterway system which allow for efficient storage, blending, and transfer of coal shipments. There is a good supply of barges and tow boats today, which provides more opportunity for increasing coal shipments.

Barge transportation is not without problems. Some locations on the upper Mississippi River above the mouth of the Ohio River, on the Ohio River's 981-mile-length from the mouth of the River at Cairo, Illinois, to its headwaters at Pittsburgh, and on certain connecting rivers, have aging lock and dam facilities. Delays occur frequently because the lock chambers have limited capacity. As a result of government funds and fuel taxes paid by barge lines, several of these locks have been rehabilitated. Since 1986, the mining industry has been joined by grain, fertilizer, chemical,

American Commercial Barge Lines, Jeffersonville, Indiana

Coal in barge.

Towboat pushing barge loaded with coal.

American Commercial Barge Lines, Jeffersonville, Indiana

CSX Transportation, Jacksonville, Florida

CSX Engine 600 pulling unit train of coal.

Underground coal miner operating Fairchild Scoop.

Fairchild International, Glen Lyn, Virginia / Grubb Photo, Bluefield, West Virginia

Tonnage of Coal Shipped on United States Waterways in 1997

Waterway	Millions of Tons
Mississippi River, Minneapolis, MN to the Gulf of Mexico	42.4
Illinois Waterway	4.9
Gulf Intracoastal Waterway	7.4
Black Warrior-Tombigbee River System (Alabama)	11.3
Tennessee-Tombigbee Waterway	2.5
Ohio River	134.2
Monongahela River (Pennsylvania)	33.7
Kanawha River	19.2
Big Sandy River (Kentucky & West Virginia)	15.9
Green & Barren Rivers (Kentucky)	6.6
Cumberland River (Kentucky & Tennessee)	10.8
Tennessee River (Kentucky, Tennessee, & Alabama)	19.4

aggregate, and petroleum product shippers to place pressure upon the government to replace or upgrade these out-dated locks and dams.

The table to the left gives us some idea of how much coal was being hauled on the major waterways in the United States in 1997.

18 wheel transport truck leaving Ivy Creek Surface Mine, April 1999, Lodestar Energy, Pikeville, Ky.

James B. Goode, Photographer

CSX Transportation, Jacksonville, Florida

CSX engine pulling unit train through coal loading facility.

COAL TRANSPORTATION BY PIPELINE

By Thomas R. Marrero, Ph.D., and John W. Wilson, Ph.D.
University of Missouri—Columbia/Rolla

In the history of transporting coal, moving coal using pipeline technology is relatively recent. There are basically three means to convert coal so that it can be suspended in water and passed through a pipe. These means are slurry, capsule (or log), and coal-water fuel (CWF). The U.S. Department of Energy has researched CWF but is not ready to release detailed information for commercial uses. In the United States, a coal slurry pipeline has been in commercial operation for 30 years. Coal transport using a unique type of hydraulic capsule pipeline has been patented and under development at the University of Missouri for several years; however, to date there is no current commercial application of the coal log pipeline (CLP).

Coal Slurry Pipeline

The Black Mesa Pipeline has been in continuous operation since 1970, transporting 4.8 million tons of coal per year from the northeastern part of Arizona to the Mohave Generating Station at Laughlin, Nevada. The pipeline is 273 miles long, 18 inches in diameter and is designed to operate at a constant throughput for 35 years at a maximum operating pressure of 1600 psi in certain lengths along the pipeline. Thus far, the Black Mesa Pipeline has transported over 100 million tons of coal over four mountain ranges, including two particularly steep areas. Along the pipeline there are three booster pump stations to provide the power needed to move the coal mixture. In the last 12 miles, before arriving at the power plant, the pipeline descends 3,000 feet and the 18-inch pipeline undergoes a size reduction to 12 inches in order to slow the excessive speed as the coal begins its sharp descent.

Prior to the Black Mesa Pipeline being constructed, a few other coal slurry pipelines operated. These were in France, Japan, the Soviet Union, and in the state of Ohio in the United States. The Ohio pipeline was 105 miles and could move 1.2 million tons of coarse coal per year. Except for the Black Mesa Pipeline, the commercial use of coal slurry pipelines has been discontinued. The Black Mesa Pipeline is now the oldest, and only, operating coal slurry pipeline in the world.

The Black Mesa Pipeline coal slurry mixture is 50% water and 50% coal. This requires grinding the raw coal into finer particles, then suspending the fine particles in water where the mixture enters the pipeline. At the pipeline outlet in Laughlin, Nevada, the water is removed by centrifuges which spin the water out much like a washing machine removes water from clothes just after they have been rinsed.

The coal slurry flowing in the pipeline causes wear, erosion/corrosion to the pipeline itself and also the other facilities of the installation. These effects are monitored and controlled. An extensive sonic monitoring program is used, as well as several other means of inspection along the length of the pipeline. Actual wear on the pipeline wall has lived up to the designer's expectations. This system can require substantial maintenance.

The adverse effects of corrosion have been minimized by using systems called "Oxygen Scavenger Inhibitors" which reduce the amount of air intake into the pipeline system.

In the event of an unscheduled shutdown, some problems can occur in attempting to restart the system. When a shutdown occurs, the coal slurry can clog the pipeline. Most all clogs are due to sending a mixture which doesn't

6" pipeline loop at University of Missouri-Rolla Experimental Mine (320' long).

University of Missouri-Rolla School of Mines and Metallurgy Department of Mining Engineering

match specifications, shutting down the system improperly, or not following correct procedures when restarting the system. Due to extensive operating experience of over 30 years at the Black Mesa Pipeline, the restart efforts in recent years have been essentially trouble-free. Additional improvements are continually being implemented into Black Mesa Pipeline components, controls, and operating procedures. This coal slurry pipeline is expected to perform well through the end of its contract period of 2005 and beyond.

Coal Log Pipeline

In comparison to a coal slurry pipeline, the coal log pipeline (CLP) has some unique features. The amount of coal to water is larger and the transport speed is much greater. The amount of coal added to the water in a CLP is dependent on the pipe diameter, and can range from 3 or 4 parts coal to one part water. Larger diameter pipelines can use even more coal to the amount of water. In a CLP, the speed is about twice that of a coal slurry. These factors reduce the unit costs (the cost-per-lb of coal transported) and make it less costly than a coal

5.4" coal logs before pipeline circulation test at University of Missouri-Rolla.

University of Missouri-Rolla School of Mines and Metallurgy Department of Mining

slurry pipeline system. This is based on the idea that the cost of the pipeline and facilities required to move the coal is about the same.

In a CLP, the coal is crushed and the particles are pressed into cylinders at the mine site. These "coal logs" can then be injected into a pipeline and moved by flowing water. These coal logs are cylindrical and do not require any other special feature to improve their ability to be transported. Coal logs are made from raw coal as it comes from the mine, with the addition of a minimum amount of material which binds or "glues" it together. The binder material is only about 1 percent of the weight that is used for making the coal logs, it enhances the heating value of the coal transported, and helps hold the log together during the time they are immersed in water. After the coal logs are pumped from the mine site through the pipeline to their destination (usually an electrical generating plant), they are easily crushed and ground to the size specifications of the power plant. The coal logs travel through the pipeline at speeds slightly less than the initial launch. The speed depends on a number of technical factors such as specific gravity (a controlled measure of the "thickness" of the mixture—more specifically a comparison of the mass of the solid material to the mass of an equal amount of liquid) and other ratios—but it can be easily figured. The decrease in pressure due to friction, drag, or other physical factors is approximately 25% more than water flowing through the same space at a similar speed. The amount of pumping needed in a CLP can be reduced by using additives such as polyethylene oxide to help reduce drag or friction.

The diameter of a coal log is usually only nine-tenths the size of the inside diameter of the pipeline. For example, a pipeline which has a 10 ft. inside diameter could accomodate a coal log up to 9 feet in diameter. The size of the coal log can vary depending on the size of the pipe. Smaller diameters of pipe require logs which are somewhat less than the nine-tenths ratio. The larger the pipeline, the closer the size of the coal log can approach being nine-tenths the diameter. The length and weight of log which can be sent also depends upon the diameter of the pipeline.

Laboratory tests indicate that satisfactory logs can be made from bituminous coal with the sole use of a binder, whereas sub-bituminous coals may require mild heating to make logs which will hold together while being transported.

In the operation of a CLP, coal logs are placed in the pipeline directly from the machine which manufactures them. Computer controlled machines feed the pipeline. The system

Coal logs after pipeline trial at University of Missouri-Rolla.

University of Missouri-Rolla School of Mines and Metallurgy Department of Mining Engineering

has a series of locks which allow trains of logs to enter the pipeline. These trains of coal logs are designed to have enough space between them to be moved by booster pump stations that permit the passage of capsules much like one train being sent after another on the same track. At the CLP outlet, the logs fall onto a screen conveyor that separates the water used in the pipeline from the coal logs. The coal logs go to a crusher and the crushed coal is then stock piled, or directed to a place where it is fed into the furnaces of the power plant.

The operation of a CLP system has been extensively tested at the University of Missouri at the Columbia and Rolla campuses. Two-inch-diameter coal logs have been transported in pipeline loops. Larger coal logs, about 5.4 inches in diameter, have also been fabricated and tested. Further tests of larger diameter logs were conducted in 1999 at a pilot plant under construction at Columbia.

In addition to the experimental studies of the science and mechanics of coal log technology, the business department has also studied the costs of this system. They have discovered that a new coal log pipeline can compete favorably with what rail companies are charging to transport an equal amount of coal in the U.S.A. The final test will come when this research team shows that a full-scale plant can be successful.

James B. Goode, Photographer

Ivy Creek Surface Mine, April 1999
Lodestar Energy, Prestonsburg, Ky.

Spradlin Branch Surface Mine, April 1999
Lodestar Energy, Prestonsburg, Ky.

James B. Goode, Photographer

REGULATION OF COAL MINING

By Bill Caylor, President
Kentucky Coal Association

Coal mining is one of the nation's most heavily regulated industries. State and federal agencies regulate both miner health and safety and the environmental aspects of coal mining. This has not always been the case.

There were two major legislative reforms instituted around a quarter-century ago that impacted coal mining. The first was the Federal Mine Safety & Health Acts of 1969 and 1977 which dealt with safety issues in coal mining. The second was the Surface Mining Control and Reclamation Act of 1977. This legislation mandated uniform national mining and reclamation practices.

These federal laws served as the force to make our nation's mines safe and to require them to operate in an environmentally responsible manner. Because of these federal laws, over the last 30 years, the reality of coal mining has quickly outpaced its initial image.

Surface Coal Mining

(Adapted from material taken from the Office of Surface Mining's Internet web site.)

The Surface Mining Control and Reclamation Act of 1977 (referred to as "SMCRA") created national environmental standards with which all coal operations must comply. This law gave the individual states primary responsibility for regulation and the federal government a supervisory or oversight role. This concept is called "primacy." Twenty-four states have chosen to have the primary responsibility of regulating coal under the provisions of SMCRA.[1]

The supervising federal agency is called the Office of Surface Mining Reclamation and Enforcement (referred to as "OSM"). OSM's 650 employees oversee the coal industry in these 24 primacy states and can cite violations if conditions warrant. OSM's current annual budget is approximately $273 million. That sum enables OSM to support the states' surface mining programs by matching their regulation and enforcement costs dollar-for-dollar. OSM also oversees the restoring of abandoned mine lands that lacked reclamation before SMCRA was signed into law.

SMCRA requires the coal operator to reclaim the land equal or better than it was prior to mining. With the exception of mountaintop removal operations, mined land must be returned to its approximate original contour. Mountain tops may be reclaimed as flat land, which creates more valuable areas for development. Before surface mining begins, coal operators must post bonds (or cash deposits) to ensure proper reclamation. These bonds cannot be released until a coal operator has demonstrated five years of consecutive successful reclamation. The land must meet certain specifics outlined in the reclamation bond application including such things as having established approved grasses, legumes, trees, or other vegetation.

As result of SMCRA, past coal mining abuses have been halted. Coal mine operators now reclaim the land as they complete the mining. Mined lands are no longer abandoned without proper reclamation. Through 1998, Kentucky had received 18 national reclamation awards. Proper reclamation creates valuable lands such as wildlife habitats, flat mountaintops, wetlands, and industrial sites where only steep hillsides once existed.

SMCRA created an Abandoned Mine Land

[1] *The 24 primary states are Alabama, Alaska, Arkansas, Colorado, Illinois, Indiana, Iowa, Kansas, Kentucky, Louisiana, Maryland, Mississippi, Missouri, Montana, New Mexico, North Dakota, Ohio, Oklahoma, Pennsylvania, Texas, Utah, Virginia, West Virginia, and Wyoming.*

Pine Branch Coal Sales Pasture Land, Chavies, Ky.

Commonwealth of Kentucky Department for Surface Mining, Reclamation, and Enforcement. Frankfort, Ky.

Fund into which coal operators contribute 35 cents per ton for surface-mined coal and 15 cents per ton for underground-mined coal.

This money is used to reclaim pre-1977 mining sites where limited reclamation occurred. Through 1996, the fund collected over $4.4 billion. As of 1999, a balance of $1.2 billion existed in this fund.

More than 19,000 acres of pre-1977 dangerous abandoned mine waste piles have been restored to productive use. Over 2.7 million linear feet of cliff-like highwalls have been eliminated. More than 20,000 dangerous abandoned portals and hazardous vertical openings have been sealed.

Prior to conducting any mining operations, the coal company is required to obtain a permit from the state regulatory authority. Permitting of a coal mine is an extensive and time-consuming process. The following background data must be gathered:

- baseline surface and ground water quality and quantity
- geologic information
- cultural, historic, and archeological resources information
- fish and wildlife resources information
- vegetative information
- analysis of water wells to determine quality and quantity
- prime farmland investigation (if appropriate)

The permitting process can require one to two years depending on the complexity of the proposed operation. A considerable amount of advance planning is essential. After the mine operator has submitted the complete permit application, it becomes public information. Copies can be reviewed at the local state regulatory office in primacy states. Public input is provided for at several stages of the permit review process.

OSM maintains a nationwide database called the Applicant/Violator System (AVS) which contains ownership/control and violation information related to coal mining operations. The AVS is used by the federal Office of Surface Mining and by state surface mining regulators to determine

Spradlin Branch Surface Mine, April 1999
Lodestar Energy, Prestonsburg, Ky.

James B. Goode, Photographer

James B. Goode, Photographer

Larry Clay feeding cattle at Pine Branch Coal Sales Surface Mine, April 29, 1995, Pine Branch Coal Sales, Chavies, Ky.

whether a permit applicant and its owners/controllers are responsible for any unsettled federal or state violations of the surface mining law, and/or for outstanding unpaid civil penalties, Abandoned Mine Land fees, or audits. If no problems are found, the applicant is eligible to receive a permit to conduct surface coal mining operations.

SMCRA requires, as an additional prerequisite for obtaining a coal mining permit, the posting of a reclamation bond which ensures that the regulatory authority will have funds to reclaim the site if there is failure to complete the reclamation plan approved in the permit.

Once the permit is issued, mining proceeds through various stages. The first stage is the removal and storage of the topsoil. Next, the rock above the coal seam (the "overburden" or "spoil") is removed and disposed of properly. After the coal is removed, the land must be reclaimed to an equal or better condition than existed prior to mining. Mining sites are inspected on a regular basis by state inspectors with random oversight

Loading coal on Cyprus Mountain Coals' reclaimed surface mine.

Cyprus Mountain Coals / Tim Cox Photographics, Norton, Virginia

by federal inspectors.

The overburden that is removed to expose the coal seam is a mixture of broken rock. If left exposed, rain water will erode the fine particles and cause sedimentation in streams and rivers. Most reclamation, except for mountaintop removal, begins by regrading the overburden to the approximate original contour of the land before mining.

Reclamation follows closely behind the mining operation and is completed in various stages as the mining progresses around the hillside. Hydro-seeding is frequently used to re-establish vegetation on reclaimed land. A blue-green liquid containing grass seed, fertilizer, and a binding material is sprayed on the topsoil to prevent erosion until plants can be established.

Controversies

Controversial issues facing the coal industry at the turn-of-the-century involve:

- A mining process called mountaintop removal, whereby the uppermost portion of the mountain is surfaced mined, creating level land. Among many issues here are the natural beauty of the land and how the the resulting level land will be used.
- The construction of large valley fills, encompassing more than 250 acres. Here the issue

Pike County Central High School after construction on reclaimed surface mine site.

Commonwealth of Kentucky Department for Surface Mining Reclamation and Enforcement, Frankfort, Kentucky

revolves around the potential of altering the biological characteristics of headwater streams.

■ The expanded use of the Lands Unsuitable for Mining Petition process to declare lands off-limits to coal mining. If the government declares land off-limits to mining, the government must pay the property owner for the loss of his minerals.

■ The limiting of air pollutant emissions from smokestacks of coal-fired electric utilities. The concerns range from the release of carbon dioxide (tied to the global warming fear) to emissions of chemicals such as sulfur dioxide, nitrogen oxides (linked to ozone pollution), and heavy metals such as mercury.

■ The concept of watershed management for impacts on streams. This affects the ability of new mining operations to be conducted in certain watersheds.

■ The disposal of coal combustion by-products (primarily coal fly ash).

Post Mining Land Use

Mining is only a temporary use of the land. When properly reclaimed, mine sites are usually suitable for all the uses they were before mining occurred. Post-mining land use changes go hand-in-hand with economic development efforts, especially in many parts of Appalachia where

Peabody Wildlife Management Area—Pre-1977 forest.

Peabody Wildlife Management Area

Peabody Wildlife Management Area—Pre-1977 forest & pond.

Peabody Wildlife Management Area

Cyprus Mountain Coals / Tim Cox Photographics, Norton, Virginia

Cyprus Mountain Coals' post mine land reclaimed for wildlife.

Cross section of 17-year-old pine trees.

much needed level land out of the flood plain is needed for this development.

Reclaimed land has been used for regional airports, correctional facilities, government facilities, schools, fish and wildlife areas, recreational areas, farms, housing developments, golf courses, shopping centers, industrial/commercial sites, and for a host of other purposes.

Technology of Post Mining Land Use

One reclamation practice of note, which is being promoted at the beginning of the new century, is called the "reforestation initiative." This method basically involves leaving the rocks and dirt used in the reclamation process in a less dense state. Minimal handling is encouraged to avoid eliminating air spaces between the material. The theory behind this method is that unconsolidated materials will greatly enhance the growth of tree roots and, therefore, the growth of trees. What the operator saves by not compacting the mine soil, he expends in better tree planting efforts. Initial results are very promising and this reclamation technique promises to be a catalyst that will encourage the planting of hardwoods normally not found on reclaimed mine land.

Another method involves the disposal of processed sewage sludge on coal mine sites, which can provide valuable nutrients for plant growth. This technique also gives sewage processing plants a disposal alternative. Other organic materials, such as sawdust and other compounds, can be distributed across the land to promote plant growth.

Abandoned underground coal mines currently provide a clean, secure, water supply reservoir for many towns in Eastern Kentucky. This water can be used to create fish farms which can support commercial ventures such as trout runs or catfish ponds. Abandoned underground coal mines could be used for growing crops which require limited light, such as mushrooms. Stocking reclaimed sites with mountain elk, buffalo and other unique species for hunting enthusiasts has already begun. The use of reclaimed coal mine lands is limited only by one's imagination.

Miner Safety

The Federal Mine Safety & Health Acts of 1969 and 1977 introduced sweeping changes in mine safety. These reforms helped promote a decline in mining fatalities. In 1932, fatalities totaled 1,192 during production of 360 million tons. In 1940, when production totaled 513 million tons, there were 1,361 fatalities. From 1969 to 1998, fatalities dropped from 163 to 29, while total coal production nearly doubled, increasing from 569 million tons to 1.1 billion tons. Although coal production in the United States has constantly grown, there has been a sharp decrease in the number of fatalities.

At present, both state and federal agencies regulate the safety aspects of coal mining. The Mine Safety and Health Administration is the federal agency primarily responsible for the health and safety of our nation's coal miners.

Most coal producing states also have agencies that regulate safety in coal mines within their boundaries. Many of the state agencies place their primary emphasis on training and licensing coal mines. Additionally, many have enforcement capabilities if violations are encountered during inspections.

James B. Goode, Photographer

Coal miners on mantrip at Miller's Creek Mine, April 1999, Lodestar Energy, Pikeville, Ky.

Miner works problem at Mine Rescue meet.

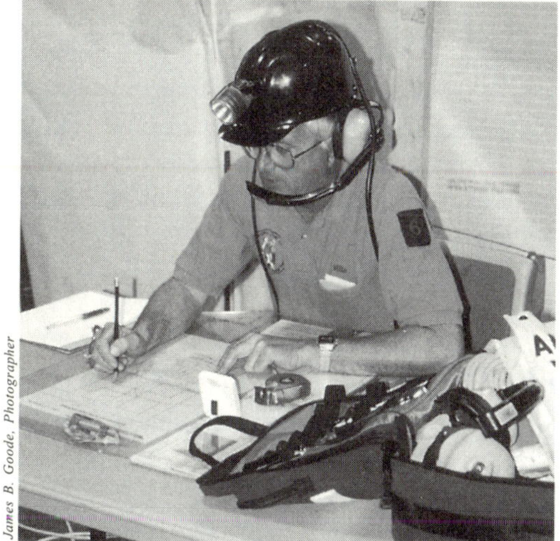

James B. Goode, Photographer

History of Mine Safety and Health Legislation

(Adapted from material appearing on MSHA's Internet web site.)

Some recognition of the dangers inherent in coal mining came at the federal level as long ago as 1865, when a bill to create a Federal Mining Bureau was introduced in the Congress. In 1891, Congress passed the first federal statute governing mine safety, marking the beginning of what was to be many subsequent years of revised and new federal legislation regulating mining activities. The 1891 law established minimum ventilation requirements at underground coal mines and prohibited operators from employing children under 12 years of age.

In 1910, following a decade in which the

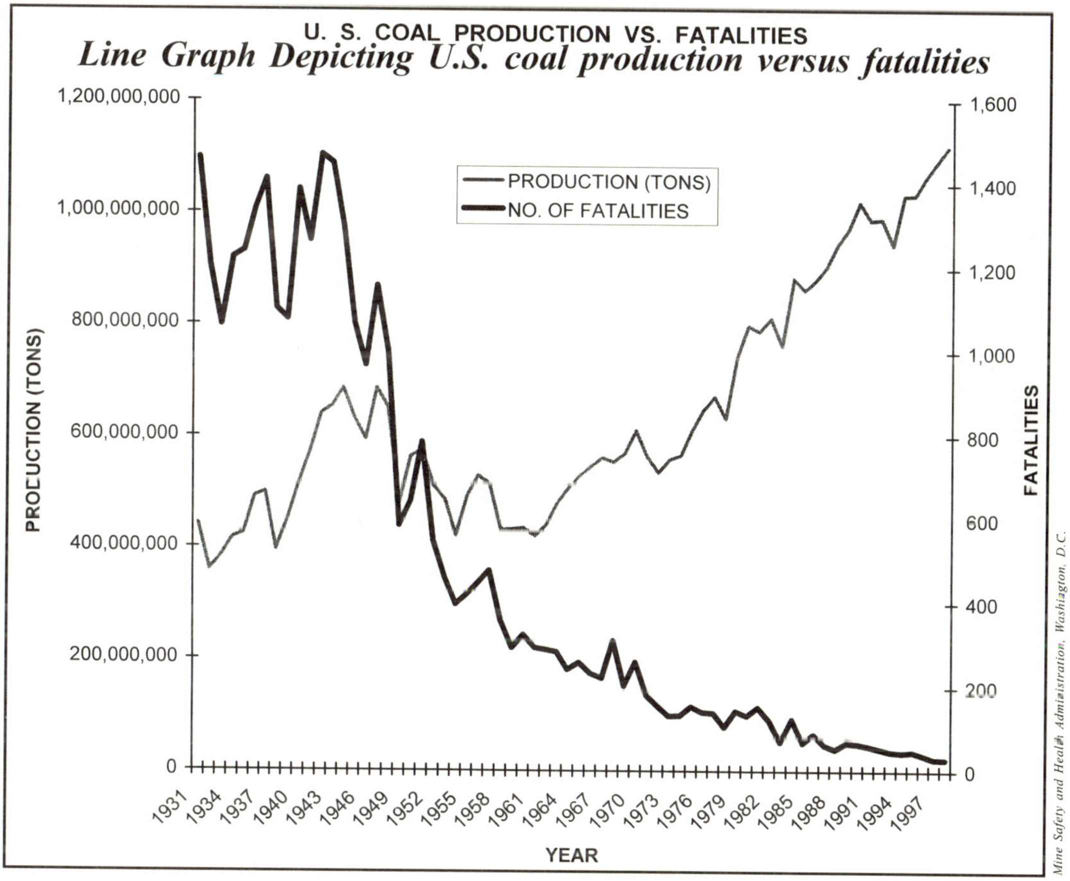

U. S. COAL PRODUCTION VS. FATALITIES

Line Graph Depicting U.S. coal production versus fatalities

PRODUCTION (TONS)
NO. OF FATALITIES

PRODUCTION (TONS): 0, 200,000,000, 400,000,000, 600,000,000, 800,000,000, 1,000,000,000, 1,200,000,000

FATALITIES: 0, 200, 400, 600, 800, 1,000, 1,200, 1,400, 1,600

YEAR: 1931, 1934, 1937, 1940, 1943, 1946, 1949, 1952, 1955, 1958, 1961, 1964, 1967, 1970, 1973, 1976, 1979, 1982, 1985, 1988, 1991, 1994, 1997

Mine Safety and Health Administration, Washington, D.C.

James B. Goode, Photographer

Miner operating shuttle car at Miller's Creek Mine, April 1999, Lodestar Energy, Pikeville, Ky.

number of coal mine fatalities exceeded 2,000 annually, Congress established the Bureau of Mines as a new agency in the Department of the Interior. The Bureau was charged with the responsibility to conduct research and to reduce accidents in the coal mining industry. They were given no inspection authority until 1941, when Congress empowered federal inspectors to enter mines. In 1947, Congress authorized the formulation of the first code of federal regulations for mine safety.

The Federal Coal Mine Safety Act of 1952 provided for annual inspections in certain underground coal mines and gave the Bureau limited enforcement authority, including power to issue violation notices and "imminent danger withdrawal orders." These were situations in which the inspector saw immediate danger and ordered the coal mine owner to withdraw his workforce from the location until the proper corrections could be made. The 1952 Act also authorized the assessment of civil penalties against mine operators for noncompliance with withdrawal orders or for refusing to give inspectors access to mine property, although no provision was made for monetary penalties for noncompliance with the safety provisions. In 1966, Congress extended coverage of the 1952 Coal Act to all underground coal mines.

The Federal Coal Mine Health and Safety Act of 1969, generally referred to as the "Coal Act," was more comprehensive and more stringent than any previous federal legislation governing the mining industry. The Coal Act encompassed surface as well as underground coal mines. Surface coal mines were required two annual inspections and four at every underground coal mines. This act dramatically increased federal enforcement powers in coal mines. The Coal Act also required monetary penalties for all violations and established criminal penalties for deliberate violations. The safety standards for all coal mines were strengthened, and health standards were adopted. The Coal Act included specific procedures for the development of improved mandatory health and safety standards and provided compensation for miners who were totally and permanently disabled by pneumoconiosis or "Black Lung" which is a progressive respiratory disease caused by the inhalation of fine coal dust.

In 1973, the Secretary of the Interior created the Mining Enforcement and Safety Administration (MESA) as a new departmental agency separate from the Bureau of Mines.[2]

MESA assumed the safety and health enforcement functions formerly carried out by the Bureau to avoid any appearance of a conflict of interest between the enforcement of mine safety and health standards and the Bureau's responsibilities for mineral resource development.

Most recently, Congress passed the Federal Mine Safety and Health Act of 1977 which is generally referred to as the "Mine Act." This is the legislation that currently governs the federal Mine Safety and Health Administration's (MSHA) activities.[3] The Mine Act amended the 1969 Coal Act in a number of significant ways and consolidated all federal health and safety regulations of the mining industry (as well as other types of mining) under a single set of laws and regulations. The Mine Act strengthened and expanded the rights of miners and enhanced the protection of miners from retaliation for exercising such rights. Under the Mine Act, fatalities dropped from 272 in 1977 to 29 in 1998. The Mine Act also transferred responsibility for carrying out its mandates from the Department of the Interior to the Department of Labor and named the new agency the Mine Safety and Health Administration (MSHA). Additionally, the Mine Act established the independent Federal Mine Safety and Health Review Commission to provide for independent review of the majority of MSHA's enforcement actions.

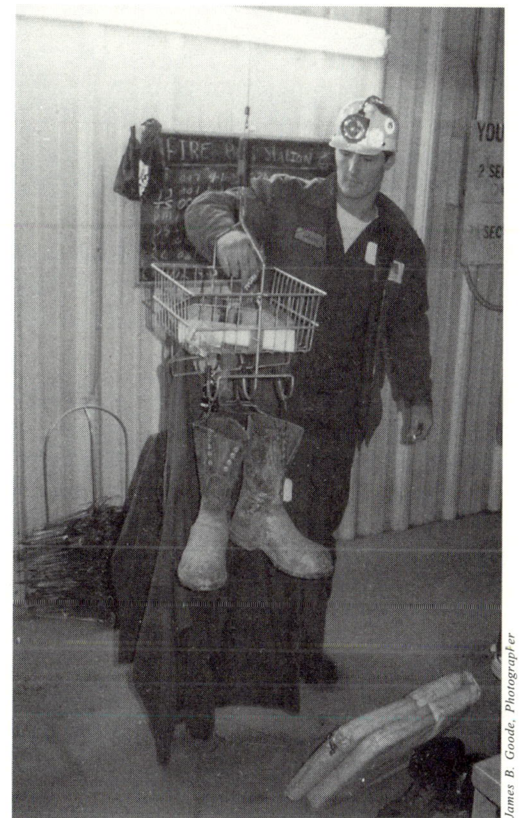

James B. Goode, Photographer

Miner in bath house at Miller's Creek Mine, April 1999, Lodestar Energy, Pikeville, Ky.

Injury Trends in Mining

(Adapted from material appearing on MSHA's Internet web site.)

Since the earliest days of mining, the job of mining coal and other useful materials from the earth has been considered one of the world's most dangerous occupations. Public concern about deaths, injuries and destruction in mine accidents has prompted passage of critical safety legislation, intensified the search for safer methods, improved training practices, and encouraged the development of more advanced technology. Growing cooperation among industry, labor and government also has contributed to making mining safer and more healthful, especially in recent years.

[2]The name of this agency was changed to Mine Safety & Health Administration (MSHA).
[3]Formerly known as the Mining Enforcement and Safety Administration (MESA)

James B. Goode, Photographer

Coal miner operating roof bolt machine at Miller's Creek Mine, April 1999, Lodestar Energy, Pikeville, Ky.

As a result of these initiatives, mining deaths and injuries have significantly declined in this century. Current low injury numbers and frequency rates are still unacceptable among safety professionals in the American mining community and the Mine Safety and Health Administration.

Total deaths in all types of U.S. mining decreased on the average during the 1900s to under 100. They reached a record low of 29 in 1998. They had averaged 1,500 or more during earlier decades. The average annual injuries to miners have also decreased steadily.

In summary, a combination of factors has been responsible for the dramatic safety gains in the U.S. mining industry since the turn-of-the-century:

■ In 1910, the U.S. Congress created the U.S. Bureau of Mines, whose primary roles were to investigate accidents, advise industry, conduct production and safety research and teach courses in accident prevention, first aid and mine rescue.

■ Federal and state laws to better advise and regulate the mining industry, to extend coverage to all types of miners, to require or encourage use of successful safety procedures and technology, to provide effective miner training, and to focus on reducing or eliminating the most serious hazards. The most far-reaching laws were the Federal Coal Mine Health and Safety Act of 1969 and the more comprehensive Federal Mine Safety and Health Act of 1977.

■ Creation in 1973 of the Mining Enforcement and Safety Administration in the Department of the Interior, which assumed safety and health enforcement responsibilities from the Bureau of Mines. And four years later, following passage of the 1977 Act, creation of the present Mine Safety and Health Administration, which was moved to the Department of Labor.

■ Introduction of vastly safer and more productive mining machines and systems, ever-safer mining methods, a growing awareness of the importance of effective accident prevention programs among both management and miners, and a more cooperative attitude toward safety issues by the mining industry, labor, and government.

The Kentucky Coal Industry at the Turn-of-the-Century

At the turn-of-the-century, the Kentucky coal mining industry has experienced historic greatness and appears to have now settled into a slow decline. Historically, the rise and fall of Kentucky's coal production has been characterized by "boom/bust" cycles. Most experts agree that the coal industry in Kentucky is entering the new century at its lowest point with a continued decline of production and price.

Coal mining is a very capital-intensive investment. What used to be an industry with many small, family-centered entrepreneurs is now dominated by large, multi-state coal companies. Coal mining is a very competitive business, which obviously attracts larger corporate entities. Coal prices at the turn-of-the-century closely match those prices paid in 1978—averaging around $24 a ton.

The trend in coal production in Kentucky shows that, although Kentucky has millions of tons of mineable coal reserves, production continues to decline. In 2000, 131.8 million tons were mined—down from a record high of 179.4 million tons in 1990. Kentucky's production is expected to continue a gradual decline. Production trends appear to be moving away from western Kentucky coal and toward coal mined underground in eastern Kentucky.

Because of the burning of coal at utility plants, electricity in Kentucky averages 4.1 cents per kilowatt-hour (the lowest in the nation). Because coal is the cheapest source of heating value, utility deregulation promises to promote the burning of coal. However favorable that may appear for the coal industry, many fear that the further reductions in electricity prices across the country will come at the expense of an already depressed coal industry.

At the turn-of-the-century, the chief fuel competitor of coal is natural gas, which burns cleaner and has no waste product. However, the instability of both its supply and price make its use an unwise choice for consumers. Regardless, actions by the U.S. Environmental Protection Agency may override common sense and force switching away from coal in the future.

Martin County Coal Corporation

Coal Preparation Plant at Martin County Coal Corporation, Inez, Ky.

THE NATIONAL ENERGY TECHNOLOGY LABORATORY (NETL)

By Douglas Deak
National Energy Technology Laboratory

The National Energy Technology Laboratory is a federally owned and operated research and development entity whose mission is to manage science and technology development projects and provide policy analyses in support of the U. S. Department of Energy's energy and environmental issues. The Laboratory has locations in Morgantown, West Virginia, and Pittsburgh, Pennsylvania. The two sites are located some 65 miles apart and are managed as a single entity. The NETL performs, procures, and partners in technical research, development, and demonstration to advance technology into the commercial marketplace, thereby benefitting the environment, contributing to U. S. employment, and advancing the position of U.S. industries in the global market.

The Laboratory was created in December 1999 through upgrading of the former Federal Energy Technology Center. The predecessor organization was formed from the consolidation of Morgantown Energy Technology Center and Pittsburgh Energy Technology Center which had been involved in fossil-energy research for the past 50 years. Approximately 550 Federal employees and 550 site support contract employees work at NETL, about half at each site.

NETL, as a Fossil Energy field organization, implements a broad range of energy and environmental programs for the U.S. Department of Energy. The nearly 700 research, development, and demonstration projects at NETL are conducted in partnership with industry, universities, not-for-profit organizations, and other national laboratories, as well as other government agencies. These "hands-on" research projects are located in all 50 states and several foreign countries. These state-of-the-art research laboratories include bench-and-pilot-scale facilities.

The NETL:

- Provides world-class fossil energy expertise
- Has an increasing nuclear-related expertise
- Provides project management expertise
- Has strong administrative, engineering, sytems analysis, and innovative, contracting skills and expertise
- Has expert ability to work with the private sector
- Has extensive research and development expertise

Major Program Areas

The National Energy Technology Laboratory is responsible for managing and implementing a broad range of programs:

- The Advanced Power Generation Systems

The Advanced Power Generation Systems Program includes both innovative combustion and gasification processes for coal and advanced turbines and fuel cells for natural gas-fueled electric-power-generating systems. A primary goal of the power program is improving efficiency. In the 21st century, today's natural gas systems will also be capable of functioning in a coal-fueled power plant. Additionally, by coupling turbines and fuel cells into hybrid systems, carbon dioxide emissions can be cut in half, compared to most typical commercial systems.

- Advanced Fuels Technology

A goal of NETL is to produce high-quality, clean-burning diesel fuels from natural gas and coal. The advanced fuels being developed will reduce particulates, nitrogen oxides, and hydrocarbon emissions. Direct liquefaction is a longer-term technology that converts coal directly into a

U.S. Department of Energy & National Energy Technology Laboratory, Morgantown, West Virginia

Aeriel view of the National Energy Technology Laboratory at Morgantown, West Virginia.

Aeriel view of the National Energy Technology Laboratory at Pittsburgh, Pennsylvania.

U.S. Department of Energy & National Energy Technology Laboratory, Morgantown, West Virginia

liquid. Coal preparation (washing or treating coal before it is used) may be the best way of removing trace contaminants or air toxics from the stack gas in emissions. Techniques under development for cleaning finely ground coal may facilitate bringing utilities in compliance with the Clean Air Act Amendments while providing valuable technology which can be exported.

■ Advanced Research and Development
Although the Advanced Research and Technology Development Program's primary emphasis is coal, several elements of the program, such as university coal research at historically black colleges and universities and other minority institutions, support all aspects of fossil energy research and development.

■ Natural Gas Exploration and Production
Building confidence in our domestic gas supply and developing low-cost production technologies helps assure that natural gas will remain an option for power generation. The Laboratory is also exploring the capture of methane, a greenhouse gas, from coal mines prior to mining and using the stored gas in applications such as fuel cells.

■ Environmental Management
NETL develops high-payoff technologies to clean up the Department of Energy's nuclear component manufacturing sites and to manage subsequent waste faster, cheaper, and more safely than currently available environmental cleanup technologies. The Laboratory can now manage nearly 100 environmental activities that provide site-specific solutions to environmental problems. The Laboratory offers cost-effective cleanup through a "virtual organization" approach by using the best and brightest from among resources in academia, private industry, and governmental agencies. The virtual organization approach is an industry-oriented model that nurtures developers toward a viable commercial product that has broad application to the nation's cleanup activities.

■ Advanced Research and
Environmental Technologies
These program activities are directed toward ensuring clean power production from coal. Higher-efficiency pollution-control systems are being developed to meet tighter environmental standards related to nitrogen oxide emissions from coal combustion. The Laboratory is exploring techniques for disposing of carbon dioxide deep in the ocean where it would remain for at least 1,000 years and not harm ocean life. The Laboratory has also studied injecting carbon dioxide into depleted oil and gas reservoirs, and through biological fixation using algae and other plants. NETL is also initiating work to sequester carbon dioxide in deep unmineable coal seams, while simultaneously recovering associated natural gas.

■ International Energy Project Implementation
NETL carries out several energy and environmental projects in foreign countries for the U.S. Agency for International Development (USAID). The Laboratory has an extensive history in bringing much-needed energy and environmental technologies to developing countries. By doing so, American technology vendors are given opportunities to penetrate new markets overseas. The Laboratory works with vendors in coordinating and implementing projects to ensure success.

International Projects

■ India
Since 1982, NETL has managed a USAID-funded project to help India use indigenous coal and biomass energy resources for electric power generation while reducing greenhouse gases. The Laboratory and USAID developed an interagency agreement for India in 1986 to promote new low-emission, low-cost, coal-based power generation systems and technologies for retrofitting older power plants for better fuel efficiency and pollution control. With NETL's help, the first South Asian pilot-scale combustion facility was built in Tiruchirapalli.

■ Poland

As a result of a 1989 cooperative agreement between Poland and the United States, the Laboratory manages a $20 million USAID project to reduce air pollution in Krakow. This two-phase program was structured to address the main sources of sulfur dioxide emissions from coal-fired utility boilers, small coal-fired furnaces, and home stoves.

■ Ukraine

A USAID-funded project allowed the Laboratory to determine the costs of repowering a power generation plant in Lugansk. This study is the focus of Ukraine's request for $200 million funding proposal to the World Bank and will allow them to no longer depend upon the unstable nuclear generation option of Chernobyl.

■ Brazil

The Laboratory is assisting the Brazilian government in creating an environmentally and fiscally sound energy policy so that domestic coal can be used to generate power as the hydro resources of the Amazon River near exhaustion. In addition, NETL organized a reverse trade mission on this topic for the U.S. Trade and Development Agency (TDA).

■ South Africa

NETL assisted the U.S. Trade and Development Agency in designing a project that examines the extent of coalbed methane deposits in the Lebowa region and assesses the feasibility of commercializing this resource for electric power generation, which may spur development in this energy deprived, economically depressed region.

■ South Korea

NETL is helping the Republic of Korea in its efforts to expand electric generation capacity to 36 gigawatts over the next 14 years. Korea is constructing 25 coal plants, 14 nuclear plants, and 12 liquid natural gas facilities. The potential market share for U.S. fossil-fueled and nuclear-generation technology is in excess of $1 billion.

The Power Systems Development Facility at Wilsonville, Alabama.

U.S. Department of Energy & National Energy Technology Laboratory, Morgantown, West Virginia

A memorandum of understanding, between NETL and two research institutes in Korea grants the Department of Energy ready access to Korean governmental and industrial energy decision makers, allowing the Department of Energy to promote the benefits of buying American technology and advice.

NETL'S Role

Given the major impact of fossil energy on the economy and the environment, the government has a legitimate role to play in establishing and implementing policy. NETL's role is to provide accurate information and advanced technology to decision makers in the administration and Congress.

■ Accurate Information

Timely data is provided on the technical, economic, and environmental implications of various policy alternatives. The Laboratory supplies accurate information on the real-cost and performance implications of trace-metal-emissions, fine particulates, nitrogen oxides, and sulphur oxides.

■ Advanced Technology

The Laboratory sets in motion rational policy and advanced technologies which need to be ready for deployment by the private sector before the government can implement a policy that requires such a technology. Given the uncertainty of the future timing and "shape" of environmental regulation, the private sector cannot accurately predict market size. Thus, private companies have little incentive to develop new technologies entirely on their own.

Introduction to Vision 21

■ Background

Widespread deregulation has caused several fundamental changes in the energy industry. While eventual effects are uncertain, the industry will clearly be far more diverse and complex. These trends lead to the vision of a future economy where power, energy, and manufacturing are closely intertwined in a highly efficient, flexible economic system. This evolution provides the foundation for Vision 21.

■ Concept

The availability of abundant, low cost fossil fuels has been a cornerstone of U.S. energy and economic strength. The continuing future use of fossil fuels in the U.S. (and quite likely, throughout much of the world) will be shaped by environmental considerations and technological innovation. The Department of Energy has reoriented its ongoing research and development program to focus on developing technologies to eliminate any environmental concerns associated with the use of fossil fuels, while assuring the continuing availability of low cost energy to fuel our economy and enhance our

U.S. Departmen. of Energy & National Energy Technology Laboratory, Morgantown, West Virginia

Candle filters for removal at high temperatures in Power Systems Development Facility at Wilsonville, Alabama.

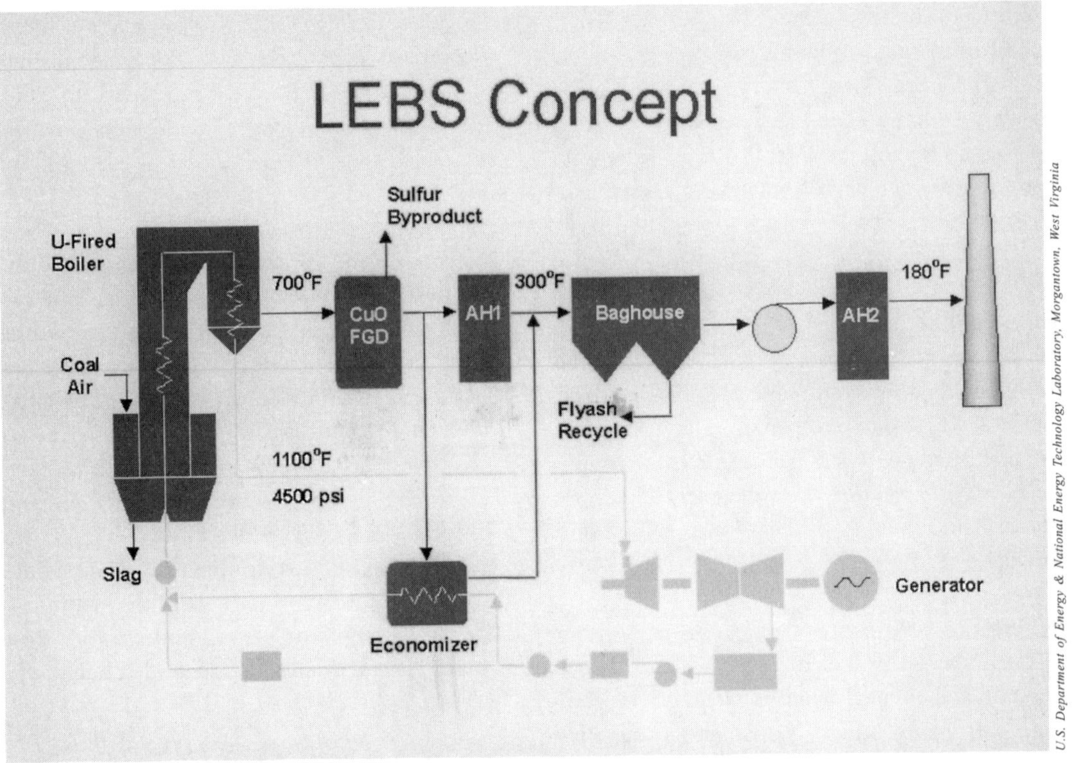

U.S. Department of Energy & National Energy Technology Laboratory, Morgantown, West Virginia

quality of life. The Department has called this program Vision 21. The Vision 21 program, which has been endorsed by the President's Committee of Advisors on Science and Technology, is a long-range, cost-shared, government-industry-academia partnership designed to produce public benefits beginning within a decade and extending well into the 21st century.

The Vision 21 program will integrate emerging concepts of high-efficiency power production and pollution controls into a new class of fuel-flexible electricity generation facilities. It will also offer a "market-entry" strategy for new concepts to produce high value fuels and chemicals from coal and other feedstocks. Discrete technology modules will offer future plant designers maximum flexibility in their choice of products, feedstocks, and environmental controls. Planners will be able to select modules according to feedstock supply and product demands of an individual region.

The Vision 21 plants will greatly reduce the energy intensity of the economy through dramatic improvements in the efficiency of electricity generation with virtually no environmental impact. Conventional pollutants could be captured and either disposed of or converted to marketable co-products. Creation of carbon dioxide and other greenhouse gases would be reduced by ultra-high efficiency technologies; then carbon emissions could be captured at the plant or offset by carbon removal processes applied elsewhere. The captured carbon could be held or potentially recycled into useful products.

At the core of a Vision 21 plant would be power generation; but rather than being limited to a single feedstock producing a single product, such a complex would be capable of processing coal, natural gas, biomass, or a combination into a range of products. In its ultimate configuration, electric power could be only one (albeit the most

U.S. Department of Energy & National Energy Technology Laboratory, Morgantown, West Virginia

Artist's view of LEBS POC Plant.

significant) of a slate of products that would also include liquid fuels and chemicals along with process heat. The exact mix of feedstocks and products would be dictated by site-specific market considerations.

■ Integration of Technologies

The integrated program path leading to Vision 21 builds on technology being developed and demonstrated today. A key component is the Clean Coal Technology Demonstration program, which establishes the foundation on which the next generation of technologies will be developed and advanced. Vision 21 would require enhanced versions of the technologies in the existing Coal and Power Systems as well as some emerging new areas to be developed within these technologies. Among the systems to be developed or upgraded are gasifiers, hot gas cleanup systems, fluidized-bed combustion, fuel cells, advanced turbines, hybrid cycles, high temperature heat exchangers, oxygen separators, hydrogen separators, fuels, and conversion processes.

■ Pre-eminent Projects

The following are three examples of ongoing NETL projects: (1) Re-powering With New Coal-Fired Technology, (2) the Power Systems Development Facility and (3) the Low Emission Boiler.

■ Power Systems Development Facility

The future of cleaner, more-efficient, coal-fired power generation rests with developing new components and intergrated systems through long-term testing and development under real conditions. This is critical to future commercialization. The Power Systems Development Facility (PSDF) project is a joint industry/Department of Energy facility operated by Southern Company Services near Wilsonville, Alabama, about 25 miles southeast of Birmingham. The objective of this project is to test and gain operating experience with new process configurations and critical components for advanced coal-based power-generating systems at a scale large enough to provide data so that they can be installed as full-sized commercial plants. Southern Company Services, Inc. is leading a team of equipment and process developers now starting up and/or operating the Power Systems Development technologies.

The ability of advanced Particulate Control Devices (PCDs) to clean hot gas is critical since they are necessary for the advanced process configurations. But more testing and development is needed to improve their mechanical integrity and reliability.

In order to arrange financing for demonstration and early commercial plants, it is necessary to demonstrate reliability and reduce the risk of many "off-the-shelf" equipment components and systems under the conditions of advanced coal conversion processes.

The PSDF is divided into two systems: a transport reactor system for parametric testing of components under typical Integrated Gasification Combined Cycle (IGCC) or Pressurized Fluidized Bed Combustion (PFBC) conditions, and an advanced PFBC system for a longer-term components in an integrated power-generation system. Combining test facilities for these technologies at the PSDF site saves more than $32 million compared with the cost of building standalone facilities.

■ Low Emission Boiler System (LEBS)

LEBS is one of the highly advanced power systems under development by a government/industry partnership. As compared to coal-fired power plants that constitute the bulk of our domestic power-generation system, LEBS offers a significantly high thermal efficiency, superior environmental performance, and a lower cost of electricity. Among the advanced power systems now being developed, LEBS is the closest to entering the commercial market and is expected to be available for commercial applications in 2002.

The LEBS program, initiated in 1992 by three industry teams, has four phases. Phase I, completed in 1994, included technical and economic evaluations of possible plant subsystems, a systems analysis of the entire power plant, and the preliminary design of a commerical-scale, 400-Mwe LEBS plant (see page 77). In Phase II, three industry teams continued engineering analysis and modeling activities and conducting experimental testing of plant subsystems at scales of approximately 3-10 Mwe. Phase III produced site-specific designs for proof-of-concept test facilities (10-80 Mwe in size) and updated commercial plant designs and economics based on Phase II results. At the end of Phase III in 1997, the Department of Energy selected one team, DB Riley, Inc., to continue to Phase IV. Phase IV, the final portion of the program, includes detailed design, construction, and operation.

UNIVERSITY OF KENTUCKY DEPARTMENT OF MINING ENGINEERING COAL-RELATED RESEARCH

By Dr. Richard J. Sweigard, Mining Engineering, The University of Kentucky

The Department of Mining Engineering at the University of Kentucky is fortunate to have a wide range of skills represented by its faculty. This expertise includes the standard areas of mining and mineral research such as mineral processing (cleaning coal of undesirable content), ground control (measures taken to prevent roof falls or coal bursting from the seam due to pressure or other factors), ventilation (venting dangerous gases and bringing fresh air to the point of mining activity), and mining systems (blending and mixing different coals to arrive at a product which meets particular requirements for its consumer). However, they also include several developing areas such as mine electrical systems, human factors, computer simulation, surface mine reclamation, and control of the environmental effects of mining. Because it has become more difficult to draw academic subject boundaries around many problems facing the coal industry, several researchers from the Department of Mining Engineering are involved in research projects that include researchers from other fields.

Dr. Daniel Tao works in the area of mineral processing and is a recent addition to the Department. Along with his colleagues at the UK Center for Applied Energy Research, he has worked on a number of research projects in the areas of fine particle processing/separation, fine coal dewatering/reconstitution, column flotation, surface chemistry, solid waste utilization, and others.

A major effort has been placed on the development of a process to create fine coal which contains less than 20% moisture. This is called a "dewatering" technique (removing water from coal). There are at least three fine coal dewatering techniques: vacuum filtration (this works like a vacuum cleaner to remove water from coal),

Dr. Richard Sweigard, Chairman of the University of Kentucky Mining Engineering Department.

centrifugal filtration (spinning the water from coal much like the spin cycle of a washing machine removes water from clothes), and hyperbaric filtration (forcing water from the coal by using extreme pressure). His studies of these techniques have revealed that fine coal dewatering is more effective when one controls the surface properties of the coal by using flocculants, surfactants, metal ions or their combinations (these are various substances which change the surface nature of coal). The most pronounced cake moisture reduction was observed with the use of anionic flocculants which lowered cake moisture below 20% when the thickness was controlled. This reduction in moisture can save coal companies a significant amount of money. If one shipped 1,000,000 tons of coal with 3% moisture, then one would be paying to ship 30,000 tons of water.

James B. Goode, Photographer

The University of Kentucky Mining and Mineral Resources Building at Lexington, Ky. December 1999

In another current research project, he is actively studying how pyrite wastes from the coal industry and gypsum wastes from utility flue gas scrubbers (cleaning devices which are placed in the exhaust of power generation facilities) can be utilized. The process under development is designed to convert both wastes into salable products such as lime, direct reduced iron (DRI), and concentrated sulfur dioxide. All these are products which can be utilized in various manufacturing processes.

Dr. Jon Yingling studies various areas of mine systems. He is currently addressing the problems related to the somewhat unpredictable characteristics of raw coal as it comes from the mine. Some of this "run of mine" coal is excellent in quality and some is high in impurities. On average, the coal generally has enough impurities that it often cannot be sold to a utility or metallurgical customer without beneficiation, a process which removes some impurities and improves the quality by mixing or "blending" different coals.

Traditional practice is to wash all of the coal in a preparation plant, which can be expensive and cause some waste of the product. When the quality of the coal can be analyzed as it enters the plant, the good quality coal can be segregated from the coal stream and bypass the coal cleaning processes, avoiding cleaning costs and raising total mine yields.

A naive strategy for coal segregation involves using analysis to determine what coal meets customer specifications and washing and/or blending the rest. This, however, is inappropriate because the *average* characteristic of such coal will greatly exceed the customer specifications. The goal is to meet the customer specifications and maximize the amount of coal that can bypass coal preparation processes. This research team has developed rules which determine which coal should be separated—which can be accomplished in cases where the coal stream has the same or similar quality during the entire period of extraction. Real coal streams, however, frequently do not have consistent quality. Consequently, several approaches have been developed that detect these inconsistencies, characterize the current nature of the process, and help make segregation decisions on the basis of these characterizations. Long-term plans include developing approaches which allow a certain quality of raw coal to be blended with clean coal. Such a strategy should enable even greater savings in coal cleaning costs.

Over a period of more than 20 years, Dr. Kot Unrug has conducted research in the area of rock mechanics (the characteristics and nature of rock) and strata control (methods of controlling the potential roof control problems which are affected by characteristics of the differing layers of rock, dirt, and minerals). Projects were carried out to measure the ability of strata in different regions of the state to withstand stress from mining. New field equipment was used to discover a method of measuring how much weight mine floors can bear. Research on coal pillar (the walls of coal left to support the roof when the primary mining is being conducted) strength was carried out in several seams. Dr. Unrug has introduced the concept of measuring the strength of a coal seam on location rather than determining strength through a coal sample for pillar calculations.

Another area of research studied the effect of weathering upon carbonaceous rocks (rocks primarily comprised of carbon, such as coal). This led to the development of a test which provides valuable information for how roof falls and coal bursts can be prevented or controlled when the removal of the coal will be over a long period of time. Currently, this method is being used by several mining companies.

Another area of interest was mine subsidence (a sinking or settling of the surface as a result of deep mining) in the mountainous Appalachian terrain. The influence on mining of the major strata commonly encountered in the eastern Kentucky coal fields has been studied. That work led to a way of predicting where "mountain cracks" (deep fissures from a settling in the strata) due to deformities may occur.

The need for a quick way of measuring the strength of rock in its natural state in strata has resulted in significant research effort. As a result, a borehole penetrometer (a measuring device) was developed which can be used to obtain a strength profile along test holes as they are being bored. The initial design has been improved by developing a miniaturized version in the laboratory that can be used in 1.5 inch diameter holes. In addition to rock strength, the penetrometer can also determine any deformities in the rock.

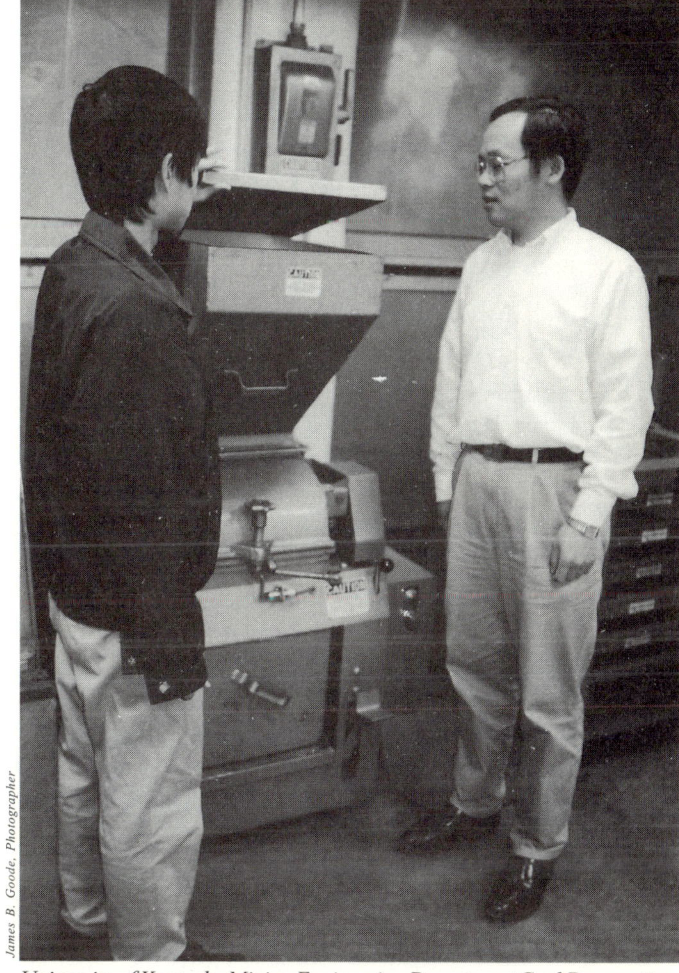

James B. Goode, Photographer

University of Kentucky Mining Engineering Department Coal Preparation Laboratory; Wendong Li, Research Assistant (Left) and Daniel Tao, Assistant Professor (Right), December 1999, Lexington, Ky.

Dr. Andrzej Wala has specialized in mine ventilation research or how to remove unwanted gases and bring fresh air to the active mining areas. More specifically, his research interests include the characteristic of how air flows in mine tunnels and in differing mining plans, how to monitor and control air flow in various mining situations, how to bring expert advice and systems

James B. Goode, Photographer

University of Kentucky Mining Engineering Department Mine Ventilation Laboratory; Jason Stoltz, Research Assistant (Left) and Andrew Wala, Professor (Right), University of Kentucky, Lexington, Ky. December 1999.

regarding ventilation to particular mining designs, how concentration of exhaust from diesel engines behaves in confined spaces, development of training exercises to prepare miners for preventing and coping with underground mine ventilation emergencies, industrial testing, and computer simulation studies of mine ventilation systems. Applying ventilation systems developed by experts in the field of managing mine ventilation and in the development of instructional materials in this area has been a productive area of research.

One of the mine ventilation research projects led to the development of a system called a Mine Ventilation Manager (MVM). This MVM system provides expert information to perform advisory consultations for underground environmental emergencies and as a link between the monitoring and responsive control aspects in the continuous maintenance of normal mine ventilation. Another application of these systems involved the development of instructional material to teach mining students and mine safety personnel the complex relationships between the mine ventilation system and a mine fire.

In another project, a procedure was developed for the external use of tracers (elements, such as non-toxic smoke, which may be introduced to air so that one can see air flow patterns) to study air flow, prior to entering the mine and taking any necessary action to control or extinguish a fire. Dr. Wala is currently researching the use of computational fluid dynamics (CFD) to assess the effects of very long walls resulting from extended cutting of coal on ventilation at the place where the coal is currently being removed.

In the area of mine electrical systems, Dr. Joseph Sottile is working on a method of detecting the early stages of electrical-component (any electrical part of a device) failure. Emphasis has been focused on electric three-phase induction motors because of their importance in the mining industry. Dr. Sottile has researched ways to record images of the patterns that voltage and electrical current make and methods of analyzing the patterns which might reveal any irregularities. This monitoring might lead to servicing or repairing the component before serious damage is realized. This approach has the advantage of not interfering with the normal operation of the equipment. This instrument design has been patented and a license agreement for selling this for commercial purposes has been obtained.

James B. Goode, Photographer

University of Kentucky Mining Engineering Department Mine Electrical Systems Laboratory; Aleck Leedy, Research Assistant (Left) and Joe Sottile, Associate Professor (Right), University of Kentucky, Lexington, Ky. December 1999.

Dr. Sottile has also studied the management of electrical energy at mining sites. This investigation addresses several aspects of energy conservation and management of electrical demand. The energy conservation component focused on reducing energy consumption by mining operations without reducing output or productivity. It involved the development of procedures for estimating how much electrical energy certain pieces of equipment might use, identifying inefficiencies, and recommending areas where reductions can be realized. Within this process, emphasis was placed on developing estimates that are easy to establish in the mining environment and recommending low cost solutions. Dr. Sottile has developed a way to forecast whether the total demand for electricity at particular facilities will exceed a certain pre-determined amount and, if so, determining a schedule for load shedding and restoration.

In addition to his recent work in Artificial Neural Networks aimed at predicting the value of the mineral based upon measurements of the seam, Dr. G. T. Lineberry has led the development of the Work Crew Performance Model (WCPM). This model represents a practical way to make sure safety is paramount during performing routine jobs which involve high-risk tasks, such as in mining. Key components of the Model include: (1) how to define a job through analysis of task, skill, and performance; (2) a method of observing workers as a way of establishing a standard by which performance and learning is measured; (3) identifying links between job competencies and measures of safety and productivity, which can then be related to costs; and, (4) providing reasons why feedback about work performance and training based on that performance can be used to assist both experienced and novice workers.

The Crew Model is founded on the simple notion that knowledge gathered from an experienced mining work force can be used to better integrate safety with production. This research seeks to improve the connection between people and technology within the mining industry. The knowledge gained from this research can be applied as solutions to other related problems that are driven by both management and workers—especially as these problems relate to the performance of work crews. The WCPM was developed originally for the mining industry with the goal of helping organizations build a higher quality performance by its work crews. This process focuses on "how" work is done, not just "what" is done or "how much" is done. This concept is now being extended into other industries.

Dr. Richard Sweigard has concentrated his research efforts in the area of reclaiming surface mine land and helping control the environmental impacts of mining. These studies include projects aimed at reducing compaction on reclaimed prime farmland soils, backfilling of auger holes in surface mine highwalls with coal combustion by-products, stabilization of coal refuse impoundments, using soil bio-engineering technology to stabilize embankments on abandoned mine land, and modeling spoil settlement on mountaintop removal sites.

In a project conducted in conjunction with researchers from the Department of Forestry, Dr. Sweigard's team is investigating new reclamation methods which employ different levels of grading to minimize how much the soil and rock is compressed which, in theory, would help hardwood trees grow better on mountaintop removal sites. The main focus of the research is to use physical characteristics of the spoil, such as how dense it is in a dry form and how resistant it would be to being penetrated by the tree roots. This research will help to predict the success of future reforestation efforts. Preliminary results indicate that there is a strong relationship between minimizing compaction and the survival rate of planted seedlings. Another objective of this project is to investigate the effectiveness of ripping the surface material to help loosen it and facilitate reforestation. Two different ripping techniques were used and the results will be compared to the other surface treatments.

Although it is impossible to quantify the total impact on the Commonwealth of Kentucky, the work done by researchers in the Department of Mining Engineering has resulted in significant gains in productivity, increased health and safety for mine workers, and a lessening of the environmental impacts of mining. The coal industry in Kentucky faces many challenges. Possibly the most serious technical challenge is the extraction of more difficult reserves as the easy-to-mine reserves of the past are being depleted. The Department of Mining Engineering is committed to developing the technology that will enable the mining industry to continue its contribution to the economy of Kentucky in the 21st Century and beyond.

COAL SCIENCE: THE FORMATION OF KENTUCKY COALS

By Stephen F. Greb
Kentucky Geological Survey
The University of Kentucky

Is coal rock or mineral? That question is more complicated than it sounds. In 1998, students from Louisville petitioned the Kentucky General Assembly to make coal the official state mineral. Legally, coal is considered a mineral resource and is taxed as a mineral. That is because any type of rock mined from the ground is taxed as a mineral resource. Rocks and minerals are not scientifically the same thing, however. A mineral is a naturally occurring solid substance with a specific chemical formula (which means the substance is always composed of the same ratio of elements and compounds) and a crystal structure. Coal does not have a crystalline structure. Therefore, coal is not scientifically a mineral, but rather is considered a rock. Rocks are naturally occurring solid substances that are composed of minerals. Coal is mostly composed of the element carbon, but also contains smaller amounts of different minerals. Because coal is deposited as layers of plant material, which are buried, it is considered a sedimentary rock.

A rock that burns

Coal is a rock that can be burned. To be considered coal, a rock must contain more than 50 percent carbonaceous (carbon) material by weight. Carbon is a naturally occurring element that is common to all organic (living) material, as are the elements hydrogen and oxygen. The three elements together form a group called hydrocarbons. Coal is an example of a solid hydrocarbon. Petroleum (oil) is a liquid hydrocarbon. Methane and other forms of natural gas are gaseous hydrocarbons. Hydrocarbons are flammable, which is why they can be used as a source of energy. Because they originated from living material, hydrocarbons, including coal, are called "fossil" fuels.

Most coal is formed from the fossil remains of plants. In fact, if one breaks open pieces of coal, he can often see imprints of fossil twigs, roots, and leaves from the original plant material. But one needs more than just a few plants to make a coal seam. It takes an overabundance of plant material accumulating for a long time, in thick deposits, and under special conditions to make a mineable coal seam.

Peat

One type of environment in which thick accumulations of plants can be deposited without being blown away, eaten, or rotting before being buried is a peat swamp. If one has ever been to a swamp or bog and had a foot sink into the black, mucky goo just beneath the surface of the water, he has seen the precursors of coal. This tangle of roots, leaves, twigs, and rotting vegetation accumulating on the floor of a swamp is called peat. Peat forms when more plant material accumulates on the swamp floor than can efficiently decay. When swamp plants die, they fall into the water and begin to decay. This means that the plant material begins to be eaten and altered by bacteria. The process of decay is most efficient when there is an ample oxygen supply for the bacteria. Water obviously contains oxygen and, in many water-logged environments, the oxygen is continuously replenished. But in swamps, the water does not circulate well and becomes stagnant. When plant material begins to decay in stagnant water, oxygen is used but not replenished. This causes the water to become oxygen poor, or anaerobic. Under anaerobic conditions, large amounts of plant material can accumulate without decaying. The accumulated, partially decayed

Plant decay and the formation of coals.

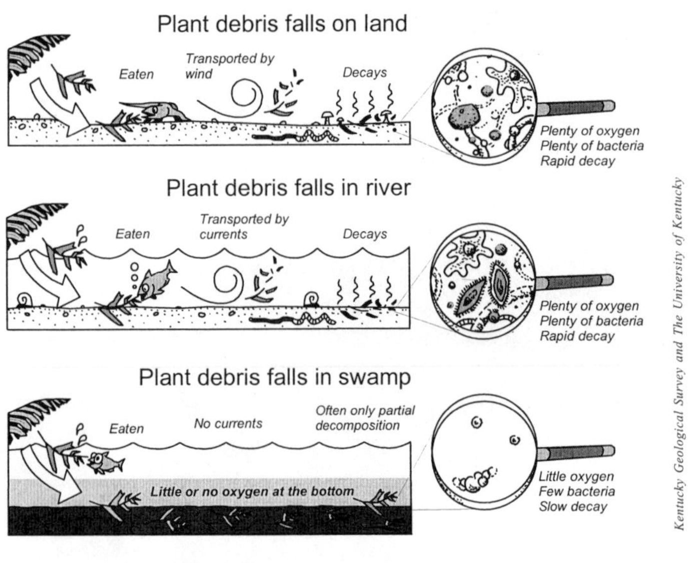

Kentucky Geological Survey and The University of Kentucky

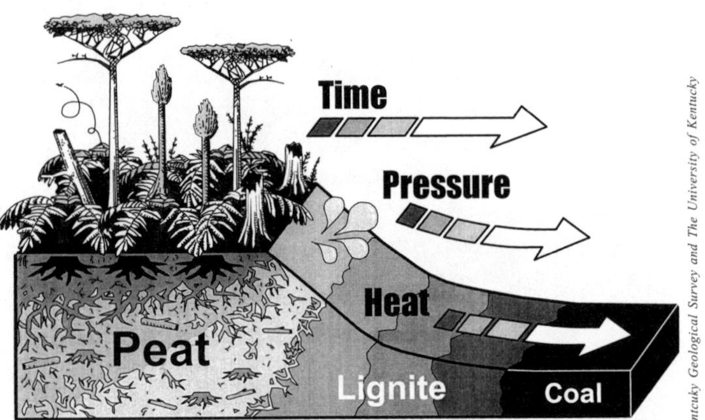

Kentucky Geological Survey and The University of Kentucky

How coal is formed.

plant material is called peat. Peat contains everything from unaltered plant parts (tree stumps, roots, pieces of bark, spores, etc.) to decayed plants, decay products, and sometimes even charcoal from swamp fires. Under the right conditions, thick peats can accumulate.

In Indonesia today, widespread swamps are common. The warm, tropical climate promotes abundant plant growth and keeps the ground wet the entire year. Under these conditions, thick peat can accumulate. In fact, some peats may be as much as 45 feet (15 meters) thick! Would this form a 45-foot-thick coal? Not necessarily. In order to become coal, the peat must be buried, compressed, and altered. It is estimated that a foot of Kentucky coal originally began as 10 feet of peat!

The process of coalification

Once the peat is buried, the combined effects of compaction, heat (from within the earth), and time begin to alter it into coal. This process is called coalification. During the early stages of coalification, the peat is compacted and water is driven out of the peat. As material is driven out of the peat, the relative amount of carbon is increasingly enriched. Differing stages along the coalification process are called rank. These ranks, or kinds of coal, in increasing order are: lignite (brown coal), sub-bituminous, bituminous, and semi-anthracite.

If coal is heated beyond anthracite rank, it becomes a form of almost pure carbon called graphite. Some people have the misconception that when coal is heated to extreme temperatures it becomes diamond, another form of pure carbon. But diamonds are formed from molten material deep within the earth, such as occurs inside volcanoes, rather than from the slow "cooking" of strata that generally occurs during coalification. When coal is heated to extreme temperatures it becomes a material that looks like pencil lead rather than diamonds.

Bituminous coal is often divided into high-volatile, medium-volatile, and low-volatile types. Volatiles are compounds within the coal that are driven out with increasing rank—hence, low-volatile bituminous coal is of slightly higher rank than high-volatile bituminous coal. As rank increases, the heating value of the coal increases. The heating value is the amount of heat produced by a specific amount of coal. Anthracite, the highest ranking coal, has the greatest value. Currently, all of the coal mined in Kentucky has bituminous rank, although there are some less altered, thin coals of lignite rank in the Jackson Purchase Region of far-western Kentucky.

Geologic periods depicting when Kentucky coals were formed.

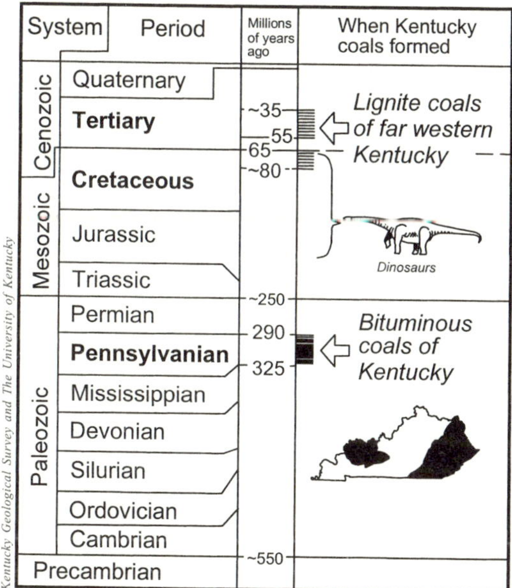

Kentucky Geological Survey and The University of Kentucky

Scientists sampling peat in an Indonesian swamp.

Kentucky Geological Survey and The University of Kentucky

Ancient peat mires

How long did it take to make Kentucky's mineable coal seams? Studies of modern Indonesian peat swamps indicate that it takes thousands, even tens of thousands of years, to accumulate peats thick enough to form a mineable coal seam. Scientists believe most of Kentucky's coals formed during a period of time called the Pennsylvanian Period, between 290 and 320 million years ago. That is a time before there were even dinosaurs. During the Pennsylvanian Period, Kentucky had a climate much warmer than it is today—a climate very similar to modern Indonesia. Also, Kentucky was a low-lying coastal area, covered with bays, swamps, wetlands, and crisscrossed by rivers. For millions of years, swamps developed and were buried, over and over again. Coal has been mined from at least 45 different seams in eastern Kentucky and at least 10 seams in western Kentucky. So, at least that many swamps were formed and were buried. The coals are separated by many different types of sedimentary rocks. These sedimentary rocks provide clues to the many ways in which each swamp was buried. In some cases, the swamps were buried by calcareous (limey) muds from the seas. In other cases, floods from ancient rivers and tidal channels buried the swamps in mud, silt, and sand.

By studying plant fossils in the coal and rocks that overlie coal, scientists can reconstruct what the ancient, coal-forming swamps looked like. The ancient swamps were composed of bushy ferns, tree-like ferns, reed-like plants called calamites (similar to large horsetail rushes), and scaley trees called lycopods. Lycopods are related to the modern club mosses, but during the Pennsylvanian Period, lycopods grew to be trees more than 100 feet (30 meters) tall. Fossil tree stumps are common in both of Kentucky's coal fields, attesting to the growth of these large trees. In underground coal mines, where these fossil stumps are found frequently, they are called "kettlebottoms" because a "kettle-shaped" end of the stump often is seen in the roof. Sometimes the animals living in the swamps or waters that buried the swamps were fossilized. Fossils of

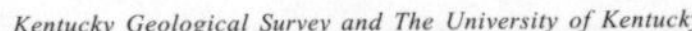

Fossil of Lycopod.

Kentucky Geological Survey and The University of Kentucky

Fossil of Annularia.

Kentucky Geological Survey and The University of Kentucky

Fossil of Alethopteris.

clams, snails, squid-like animals called cephalopods, and even fish have been found in the coal-bearing rocks of Kentucky.

Over time, layers of sediments were piled on top of previous layers of sediments that buried the peats. As the sediments were buried, they turned into rock. Calcareous muds became limestone, silica-rich muds became shale, silty sediments became siltstone, and sandy sediments became sandstone. The buried peats were "slow-cooked" beneath thousands of feet of this rock and eventually became coal.

Over time, more changes to the land occurred. The 190-mile-long Pine Mountain in eastern Kentucky was thrust upward during the formation of the Appalachian Mountains. Rivers and streams began to slowly erode and carve valleys into the thousands of feet of material that buried the peats. For millions of years, this erosion and weathering cut away at the landscape, until finally the coal beds were exposed as we see them today.

The lignite coals of the Jackson Purchase have a different history. They formed during the late Cretaceous (65 to 80 million years ago) and early Tertiary periods (35 to 55 million years ago), well after the formation of the coals from the eastern and western Kentucky coal fields, and after formation of the Appalachian Mountains. During this time, sea level rose and fell several times. When sea level was high, it drowned much of the Mississippi Valley, and swamps formed. In the western United States, thick peats formed. In Kentucky, only thin peats of limited extent accumulated. Dinosaur footprints have been found in Cretaceous coal in the western United States. We now know that dinosaurs walked across the ancient lignite swamps of Kentucky, because of a recent discovery of a dinosaur fossil in Perry County, Kentucky.

The Cretaceous and Tertiary lignite-forming peats of the Jackson Purchase Region were deposited after many of the events that shaped the earlier peats into the coals of the eastern and

Kentucky Geological Survey and The University of Kentucky

Fossil of Harlan County tree stump and roots.

western Kentucky coal fields. They were not buried deeply—and for only 1/4 to 1/9 of the time (depending on the bed) than the mineable coals of the eastern and western Kentucky coal fields. In essence, the lignite coal did not "cook" as long as the coals from the other coal fields, which is why they have less rank. None of these lignite coals is currently mined.

Formation of Kentucky's two coal fields

The geologic map shows the distribution of rocks at the surface in Kentucky. Coal-bearing rocks occur in the eastern and western parts of the state, in the group of rocks labeled Pennsylvanian strata. The reason the two areas are separated has to do with geologic processes that have changed the surface of Kentucky through the ages. If one were able to take a knife and slice through the surface, one would see how rocks beneath the surface of the state are bent upwards and downwards across great distances. The samples of strata extracted from drilling for oil, gas, and water have often given geologists the information they need to correlate or trace rock layers beneath the surface. Geologists have determined that rocks at the surface in the center of the state occur thousands of feet beneath the

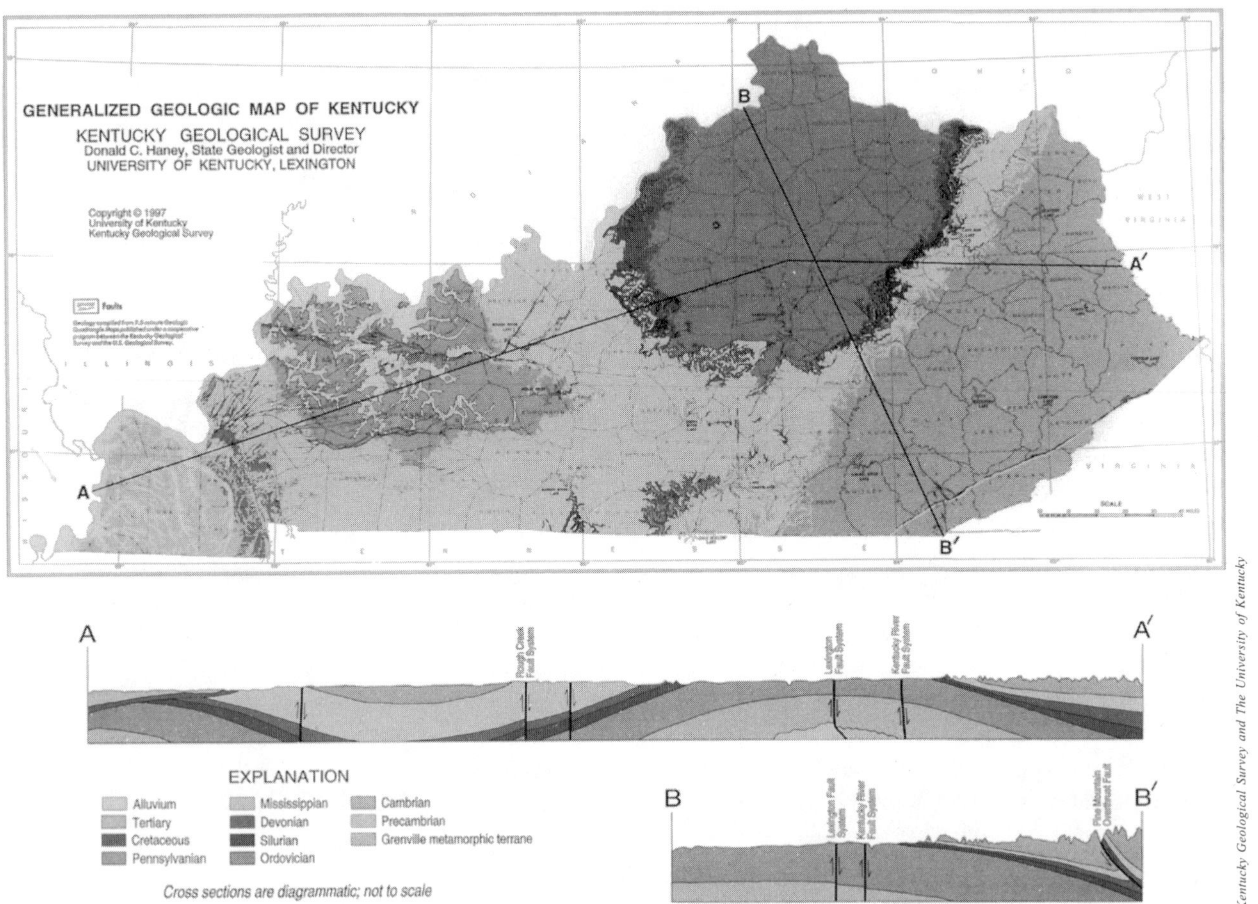

Kentucky Geological Survey and The University of Kentucky

surface in the eastern and western parts of the state. Likewise, strata in the eastern and western parts of the state are missing in the center of Kentucky. The upward-bending strata in the center of the state forms a structure called an arch.

Strata bending downward into bowl-shaped depressions on either side of the arch form structures called basins. The basin in the east is called the Appalachian Basin, and it continues into Pennsylvania, Ohio, West Virginia, Tennessee, Virginia, Georgia, and Mississippi. The basin in the west is called the Illinois Basin, and it continues into Illinois and Indiana. Pennsylvanian strata, which contain the mineable coals of Kentucky, are confined to these two basins.

Differences in Kentucky coals

Because coals are the fossil remnants of swamps, they are extremely variable in composition. If one has ever seen a swamp on television, or actually visited a swamp, he or she could see how some of the differences might occur. There are areas of swamps in which thick forests occur; in other areas, reeds dominate; in still others, there is open water. Hence, there are variations in the types of plant materials that accumulate in different parts of a swamp. Also, sediment is carried in waters around peat swamps. During floods, that sediment is deposited on top of the peat. If peat continues to be deposited after the flood, then the sediment may be mixed into the peat or be left as a sediment layer in the peat. During burial, more sediment is deposited on top of the peat. As the peat is compacted, some of this sediment may be mixed into the upper part of the peat. Sediment is mostly composed of minerals that will not burn. It is called "ash" in coal. The higher percentage of ash in coal, the lower the percentage of material that will burn. In Kentucky, mined seams have average ash contents between 8 and 12 percent. The best coals have little ash.

One of the most common and important minerals in coal is composed of iron and sulfur, and is called pyrite, or "fool's gold." Pyrite can form from sulfur that is in the plants that make up the peat, and from sulfur transported into coal during coalification by water in the ground. When coal is burned, the pyrite breaks down, and sulfur is released into the atmosphere as sulfur dioxide ($S0_2$). Some coal beds contain high amounts of sulfur; others do not. For various reasons, most of the coals mined in western Kentucky have sulfur contents of 3 to 5 percent. Eastern Kentucky coals often have sulfur contents of less than 1.5 percent. Sulfur occurs in coal in three forms—organic combinations in the coal, sulfides (generally pyrites or marcasites), and sulfate. The Federal government regulates how much sulfur dioxide can be released into the atmosphere when coal is burned in power plants to create steam for generating electricity. Hence, in today's marketplace, coals that contain lower sulfur amounts are generally considered more valuable than those with higher sulfur amounts.

Coals that contain more sulfur than allowed by law can be blended with coals that contain very little sulfur or can undergo a process to chemically and physically remove some of the sulfur. The process is called "beneficiation" or "washing." Sulfur can also be removed from the smoke formed when coal is burned, by using "scrubbers" on power-plant smoke stacks. These scrubbers make the sulfur combine with powdered lime to form a new mineral called gypsum. The gypsum is a solid and falls into a collecting chamber at the bottom of the stacks, rather than being released as a gas into the atmosphere. Nearly half of the power plants in Kentucky are equipped with scrubbers, which is significant since the national average is only 22 %.

Differences in coals around the United States

Not only are each of the coals in Kentucky different, and not only are average coal compositions from Kentucky's two coal fields different, but coals mined across the United States are also different. Some coals differ in rank. For example, coals mined in the Appalachian Mountains of Pennsylvania have undergone more heating and alteration because they were affected by mountain-building processes. These coals have anthracite rank. Anthracite coal produces more heat when burned than coals of lesser rank. The heating value of coal is generally measured in British Thermal Units, or BTUs. The heating value of coal is the amount of BTUs produced by burning a pound of coal. When anthracite is burned, it has a heating value of more than 15,000 BTU/pound. In contrast, lignite often burns at less than 8,500 BTU/pound. Kentucky's mined coals are high-volatile bituminous rank and have average heating values of between 11,000 and 13,000 BTU/pound.

In recent years, Wyoming passed Kentucky and West Virginia to become the leading annual producer of coal in the United States. One of the advantages of Kentucky coal is that it is of higher quality and takes less volume to produce the same amount of heat as Wyoming coal. Wyoming coals are often very thick, with some seams being over 40 feet (12.5 meters). The average height of Kentucky coals is between 3 and 5 feet (< 2 meters). Wyoming coals have very low sulfur contents—generally less than 0.5 percent. The low sulfur contents are one of the main reasons that Wyoming coal has become increasingly valuable in today's marketplace.

To learn more about the science of coal, one should visit the Kentucky Geological Survey web site at the University of Kentucky in Lexington, Kentucky:
http://www.uky.edu/KGS/
Or see the following references:

Cobb, J. C., et al. *Kentucky Coal*. Lexington, Kentucky: Kentucky Geological Survey, Ser. 11, Reprint 20, 1985.

Cobb, J. C. and C. F. Eble. *Sulfur In Kentucky Coal and the Clean Air Act Amendments of 1990*. Lexington, Kentucky: Kentucky Geological Survey, Ser. 11, Information Circular 38, 1992.

Cobb, J. C. et al, eds. *Coal and Coal-Bearing Rocks of Eastern Kentucky*. Lexington, Kentucky: Kentucky Geological Survey, Guidebook and Roadlog for Coal Division of Geological Society of America, Field Trip No. 14, 1981.

Eble, C. F. and J. C. Cobb. *Trace Elements In Coal—The Next Challenge*. Lexington, Kentucky: Kentucky Geological Survey, Ser. 1, Information Circular 48, 1994.

Greb, S. F., D. A. Williams, and A. D. Williamson. *Geology and Stratigraphy of the Western Kentucky Coal Field*. Lexington, Kentucky: Kentucky Geological Survey, Ser.11, Special Pubication 8, 1992.

McGrain, Preston. *The Geologic Story of Kentucky*. Lexington, Kentucky: Kentucky Geological Survey, Ser. 11, Special Publication 8, 1983.

Weisenfluh, G. A., et al. *Kentucky's Coal Industry—Historical Trends and Future Opportunities*. Lexington, Kentucky: Kentucky Geological Survey, Ser. 11, Information Circular 59, 1998.

United States Steel Longwall, mine #37, Cumberland, Ky., 1980s

James B. Goode, Photographer

UNIVERSITY OF KENTUCKY CENTER FOR APPLIED ENERGY RESEARCH (CAER)

By Ari Geertsma, Ph.D., Director
Center for Applied Energy Research

The University of Kentucky Center for Applied Energy Research (CAER) is well positioned to be a key research and development institution with the resurgent interest in coal as a primary fossil fuel for the supply of electricity, not only for Kentucky, but also for export nationwide. Located in Lexington, it is a state-mandated research center that was established in response to the energy troubles of the 1970s. Management control passed on to the University of Kentucky in 1988. Historically, the center played a significant role in the state's energy policies, especially in the days of direct coal conversion and oil shale research and development. Subsequently the Center also moved into fuels and chemical materials.

Currently, CAER has an annual expenditure of $5.67 million, of which 40% is external money, mostly from the Department of Energy. The Center strives to maintain a 50/50 balance between state (University of Kentucky) funds and extramural funds.

The Center has a very capable team with an excellent international reputation in fuel science and energy research. The approximate 100 employees include roughly 35 scientists and engineers, 25 undergraduate and graduate students, 6 postdoctoral researchers, with the remainder being support staff.

Goals and Strategies

The broad goal of CAER is to excel as a research and development center with an international reputation, focusing on the optimal use of Kentucky's energy resources for the benefit of its people. In order to achieve this, close collabo-

The University of Kentucky Center for Applied Energy Research.

Dr. BK Parekh with column flotation technology. This techology separates ash from clean coal prior to combustion and was commercialized about a decade ago for use by coal preparation plants.

ration with industries and with State and Federal funding agencies is pursued. The CAER staff maintains and grows interactions with UK faculty with a view to expand programs and maintain strong communications with the University. Along this line, the CAER is increasingly becoming a center preferred by students for graduate studies. The staff actively participates in academic programs at the university.

As one of the University of Kentucky's 13 research centers, the staff members perform sound fundamental and applied research to develop relevant industrial technologies. Researchers collaborate with external stakeholders to implement these technologies. The contribution to the education and training aim of UK is satisfied in-part by participating in the education of students at all levels and helping to contribute to labor-force development in Kentucky. Much of the current work is environmental in nature. By providing technologies to improve the environment, and the use of coal, CAER enhances the quality of life within the Commonwealth.

Work Groups

■ Environmental and Coal Technologies

This group is at the forefront of coal beneficiation and utilization technologies through research and development of new avenues to add value to Kentucky coal and proposes viable technologies for industrial application. Major issues include 1) coal clean-up 2) wet and dry beneficiation 3) waste disposal and utilization 4) fly ash 5) coal combustion and gasification 6) alternative feedstocks for combustion and gasification 7) effluents and water pollution in the coal industry 8) land remediation and technologies to reduce the emission of undesirable substances like nitrogen oxide, sulfur dioxide and mercury

■ Clean Fuels and Chemicals

This group emphasizes the use of catalytic pro cesses for the production of synthesis gas derived hydrocarbons. Major issues include 1) catalysis of the production of synthesis gas and the use of synthesis gas for the production of ultra clean diesel and petrochemicals 2) catalyst separation from products as well as the detailed characterization of catalysts and products 3) researching and modeling engineering aspects of the reactor systems. The Catalysis Research and Testing Center also forms part of this group. The focal point of the testing center, housed within the CAER, is to define scientific principles involved in catalysis as well as applying these scientific principles to the many applications of catalysts in industry.

Coal liquefaction lab. CAER researches the processes that turn coal into a transportation liquid.

■ Carbon Materials

This group develops new carbon materials and technologies to address energy and environmental needs. Emphasis is placed on the larger-scale production of these materials and the transfer of resulting technologies to industry. It will advance the scientific understanding of these materials through fundamental and applied research. Carbon materials derived from fossil resources remains a core activity, with particular emphasis on high-value carbon materials derived from Kentucky coals.

■ Analytical Services

The CAER has established an industrial support service whose goal is to apply the Center's expertise and facilities to the solutions of industrial problems. The state of the art analytical equipment and an experienced staff combine to provide a unique problem-solving capability with a proven track record. These resources are made available to industrial, academic, and government entities to address technical problems ranging from the routine to the advancement of the state-of-the-art. In addition to supplying quick answers to well-defined problems, the Center collaborates on longer-term efforts for applied research, prototype development, and potential commercial applications.

The CAER constantly strives to meet these important goals, in order to help support the university in becoming a Top 20 research university. Future work and collaboration is constantly in the planning and follow-through stages. For more information on the center's work, services, and potential collaborations, see the web site: www.caer.uky.edu.

GLOSSARY

AC
Alternating Current.

Active Workings
Any place in a mine where active mining is being conducted or where miners are required to travel or work.

Adit
A nearly horizontal entrance passage which does not exit the other side of the mountain or hill.

Advance Mining
During the initial mining of coal, all extraction moves forward as far as possible toward the property boundaries.

AFT
Advanced Fuel Technology.

Legend of Abbreviations

AFT	Advanced Fuel Technology
ANFO	Ammonium Nitrate/Fuel Oil
ANN	Artificial Neural Networks
APFBC	Advanced Circulating Pressurized Fluidized Bed Combustion
AVS	Applicant/Violator System
B.T.U.	British Thermal Unit
CFD	Computational Fluid Dynamics
CLP	Coal Log Pipeline
CWF	Coal-Water Fuel
FMB	Federal Mining Bureau
GPS	Global Positioning System
IGCC	Integrated Gasification Combined Cycle
MESA	Mining Enforcement & Safety Adminstration
MSHA	Mine Safety Health Administration
MVM	Mine Ventilation Manager
NETL	National Energy Technology Laboratory
OSM	Office of Surface Mining Reclamation and Enforcement
PCD	Particulate Control Devices
PSDF	Power Systems Development Facility
SMCRA	Surface Mining Control and Reclamation Act of 1977
WCPM	Work Crew Performance Model

Aggregate Flotation
A clean coal technology that involves floating chemically conditioned, finely-ground coal in water. As the coal sticks to the rising air bubbles, nearly all of the inorganic sulfur sinks to the bottom of the mixing tank. The result is a cleaner burning coal that emits less sulfur dioxide.

Air Course
A passage or tunnel in a mine through which air is circulated.

Anemometer
An instrument equipped to measure air velocity.

ANFO
Ammonium nitrate/fuel oil mixtures used for blasting.

ANN
Artificial Neural Networks is a system developed by the University of Kentucky to predict the value of coal based upon certain specific measurements of the seam.

Anthracite
This is the hardest coal. It is almost pure carbon and has the highest BTU content. Scientifically defined as coal with a volatile ratio equal to 0.12 or less.

AOC
This is a term applied to surface mining and signifies that after mining, the earth must be returned to its approximate original contour or state.

APFBC
Advanced Circulating Pressurized Fluidized Bed Combustion Combined Cycle is a process used to repower existing steam turbine electrical generation units to increase the amount of electricity generated.

Area Mining
Conducted over large areas of land and practiced on relatively flat terrain, this type of mining employs giant drag lines and shovels which strip away the overburden above the coal. In the mountains, a modified version of area mining called Mountain Top Removal is practiced. Usually, smaller machines prepare the mountain site for the larger shovels and drag lines.

Atmospheric Fluidized Bed Combustion
AFBC is a type of clean coal technology that uses pulverized coal mixed with limestone. The limestone absorbs sulfur from the gases resulting from the combustion of the coal. The process reduces approximately 90 percent of sulfur dioxide emissions and causes a moderate reduction of nitrogen oxide emissions.

Auger
A horizontal rotary drill that utilizes a screw configuration to penetrate, break, and then transport drilled material. Auger-type devices are popular in soft coal.

AVS
Applicant/Violator System is a computer database developed by OSM which contains ownership control and violation information related to coal mining operations. This information can be used to determine whether a permit applicant or the owners/controllers are responsible for any unsettled federal or state violations of the surface mining law, and/or for outstanding unpaid civil penalties, Abandoned Mine Land fees, or audits.

Backfill
The dirt and rock removed to access the coal bed and then used to refill the mined area or for filling in hollows.

Barrier
Any device which is used to bar or keep something out of another place. In mining, usually a block of coal.

Barrier Pillars
Solid blocks of coal left between sections of a mine to protect the tunnels and coal headings from the crushing weight of rock, slate, and dirt overburden.

Bearing Plate
A steel plate used to distribute a given load. In roof bolting, the plate used between the bolt head and the roof of the mine.

Bituminous
A soft coal with the second highest BTU content. It is the most common coal found in the United States and is, generally, used to generate electricity and to make coke for the steel industry.

Black Damp
Usually refers to carbon dioxide ($C0_2$). More exactly, this is a mixture of carbon dioxide and nitrogen. Sometimes refers to an oxygen depleted atmosphere.

Blast
Refers to a controlled explosion which is used to loosen material to aid in mining coal.

Blasting Agent
Refers to any material consisting of a mixture of fuel and an oxidizer which cannot be detonated by a No.8 blasting cap or be classified as an explosive.

Blasting Cap
A device containing a small charge of detonating compound which is ignited by electric current or a spark which then triggers a larger explosion.

Bond
Cash deposits or insurance which is held by a government agency until a coal company meets a pre-determined agreement—usually regarding specifics of a reclamation plan.

Brattice
A fire resistant fabric or plastic canvas which is used to partition any mine passage for purposes of confining air and directing it into areas where work is occurring in the mine.

Breakthrough
A ventilation passage which is cut through a pillar of coal to connect two tunnels or rooms.

Breast Auger
An out-moded, hand-operated, screw-like bit which is used to drill into coal to create a hole for blasting charge. The drill is shaped like a brace-and-bit and fits into a metal plate mounted on the miners' chest.

Brogan
A very durable leather shoe or boot made for practical purposes rather than for looks.

B.T.U.
British Thermal Unit. This is a measure of energy required to raise the temperature of one pound of water by one degree Fahrenheit.

Bug Dust
Fine particles of coal or other material resulting from the boring or cutting of the coal face by a drill machine.

Bump or Burst
A violent dislocation of coal in a mine working which is caused by extremely high stress in the surrounding rock.

Bureau of Mining
A former agency established by Congress in 1910 as a new agency in the Department of the Interior and charged with conducting research and reducing accidents in the coal mining industry.

Butt Entry
Usually refers to an entry which is at a right angle to the face.

Canaries
In the early part of the century, these small, delicate birds were carried into the coal mines to test for noxious air. Because of their size and respiration rates, they would succumb before the air became deadly for the miner.

Cannel Coal
A fine, even grain, non-caking coal which contains a high percentage of hydrogen, burns with a long, yellow flame, and is extremely easy to ignite.

Canopy
A protective roof found on various kinds of mining machinery.

Cap
A miner's safety helmet. Also refers to a highly sensitive encapsulated explosive that is used to detonate a larger, less sensitive explosive material.

Carbide Bit
A cutting bit made from cemented tungsten carbide. A piece of molded tungsten carbide is inserted into the end of a steel bit shank to form the bit.

Carboniferous
Any substance having a high content of carbon.

Cast
In surface mining, when the overburden is projected (usually from a blast) and thrown from above the coal into the previously mined area.

Centrifugal Filtration
The process of removing water from coal by using a device that spins the coal at a predetermined velocity and separates the lighter water from the heavier coal.

CFD
Computational Fluid Dynamics is a system developed by the University of Kentucky to measure the effects of long wall faces on ventilation.

Chain Conveyor
A device used to move coal along solid paps (troughs) by scraper crossbars attached to power driven chains.

Chock
A large hydraulic jack used to support a mine roof in long wall and short wall mining.

Circulating Fluidized Bed Combustion
A clean coal technology process that produces a mixture of coal and limestone in a liquid state by using vertically moving air. The process effectively removes sulfur and nitrogen from coal, thus reducing sulfur dioxide and nitrogen oxide from coal-burning emissions.

Clean Coal Technology

Processes developed to clean the content of coal burning emissions including sulfur dioxide, nitrogen oxide, carbon dioxide, air toxins and particulates, while enabling coal-burning facilities to meet or exceed emission standards.

Cleat

The vertical cleavage of a coal seam. This is the main set of vertical joints along which coal breaks when mined.

CLP

Coal Log Pipeline. This is a coal transport system which involves the crushing and forming of coal into cylinders or logs and propelling them with water to a destination.

Coal Barge

A coal transport vessel which is used on water and is usually moved in groups by a towboat. These vessels hold between 1,000 and 1,500 tons of coal.

Coal Camp
Town built by a coal company to house coal miners and their families. Many of these included shopping areas, entertainment facilities, school buildings, hospitals, and other basic infrastructure. There were over 20,000 coal camps in the U.S. built during a period from 1900 to 1960.

Coal Dust
Particles of coal which can pass through a No. 20 sieve.

Coal Gasification
The conversion of coal into a gaseous fuel.

Coal Mine
Area of land from which coal is extracted by surface or underground methods.

Coal Reserves
Measured tonnages of coal that have been calculated to occur in a coal seam within a particular property.

Coal Seam
A deposit of coal— usually in a layer much like the appearance of icing between layers of a cake.

Coal Slurry
1) A transportation process which requires the grinding of raw coal into small particles and suspending them in water for the purpose of transporting the compound through a pipeline to a destination. 2) Extra fine coal, water, dirt, and rock pumped from washing coal at a preparation plant into a refuse area.

Coal Washing
The process of separating undesirable materials from coal based on differences in densities.

Coke
A hard, dry, carbon substance produced by heating coal to a very high temperature in the absence of air.

Continuous Miner
A machine that eliminates the cycles of cutting, drilling, and shooting by tearing the coal from the face with a rotating drum of solid, steel bits and loading it continuously. This is differentiated from a conventional, or cyclic, unit which must stop the extraction process in order for loading to commence.

Contour Mining
Surface mining along an imaginary line which connects all points on a surface having the same elevation.

Conventional Mining
An older system of mining which uses the cyclical operations of cutting, drilling, shooting, and loading. This process is older than continuous mining.

Conveyor Belt
A motor-driven continuous (looped) belt made of reinforced rubber or rubber-like substance on which coal and other materials are carried.

Crib
A roof support of wooden prop timbers or ties, laid in alternate cross layers, log-cabin style. Sometimes the interior space is filled with debris.

Crossbars
Any horizontal support placed between two vertical supports. Can be made of various materials including wood and metal.

Crosscut
A passageway driven between the entry and its parallel air course or air courses for ventilation purposes. A crosscut can also be a tunnel driven from one seam to another through or across the intervening measures.

Crusher
A machine used for crushing rock or coal. Some types include: 1) Ball Mill 2) Gyratory 3) Handsel Mill 4) Hammer Mill 5) Jaw Crusher 6) Rod Mill 7) Rolls 8) Stamp Mill 9) Tube Mill.

Cut
To mechanically slice a coal seam.

Cutting Machine
A generally outmoded machine that will cut a four to six inch slot in the coal (usually at the base of the seam) which allows the coal to fall and break when blasted or picked from the bed.

CWF
Coal-Water Fuel.

Demonstrated Coal Reserves
Coal deposits that are potentially mineable on an economic basis with existing technology. The U.S. Department of Energy estimates that currently there are about 500 billion tons of demonstrated coal reserves in the United States.

Detectors
Specialized chemical or electronic instruments used to determine the presence and qualities of gases.

Detonators
Devices containing a small charge that is used to start a larger explosion. May refer to blasting caps, electronic detonators, and delay electric blasting caps.

Dewatering
Removing water from coal.

Distribution Box
A portable apparatus with an enclosure through which electric circuit is carried to one or more outgoing cables from a single incoming feed line.

Dragline
A large powered machine which drags a high capacity bucket mounted on a cable to remove overburden from above a coal seam at a surface mine.

Draw Rock or Slate
A soft slate, shale, or rock from 2" to 24" thick located immediately above certain seams which falls quite easily when the coal is withdrawn.

Drift Mouth
An entry into a coal seam at surface level. Generally located on the slope of a hill where the mine then advances horizontally into the coal seam.

Duckbill
An outmoded type of shaker conveyor/loader placed under a coal cutting machine.

Dummy
In blasting, a paper bag filled with sand, clay, or other inert material which serves as stemming for a charged hole.

Entry
An underground passage used for haulage, ventilation, or as a manway. This is the place where coal was extracted from the seam in the initial mining.

Face
The principle operating place in a mine where fresh coal is being exposed and extracted. A mine may have many operating faces.

Face Conveyor
Any conveyor used parallel to a working face which delivers coal to another conveyor or transport car.

Fall
When a mass of roof rock, slate, and/or coal collapses to the floor of the mine.

Fault
A slip-surface between two portions of the Earth's surface that have moved relative to each other. A fault is a result of high Earth stresses and generally creates a surface failure.

Federal Mining Bureau
Established by Congress in 1865 to address the dangers of coal mining. The functions of this agency were assumed by three agencies including National Institute of Occupational Safety and Health, Mine Safety and Health Administration, and Department of Energy.

Feeder
A machine that evenly feeds coal onto a conveyor belt.

Fill
Any material that is put back in place of the extracted ore.

Flame Safety Lamp
A lamp used in coal mines to detect the presence of methane gas or to indicate oxygen deficiencies. This lamp is equipped with steel wire gauze which prevents the passage of the flame within the lamp to the outside and igniting explosive gas.

Flue Gas Desulfurization
Any of several forms of chemical/physical processes that remove sulfur compounds formed during coal combustion. Scrubbers are often used to combine the sulfur in gaseous emissions with another chemical medium to form inert sludge which must then be removed for disposal.

Fluidized Bed Combustion
A clean coal technology process that removes sulfur from coal during combustion. In a fluidized bed boiler, crushed coal and limestone are suspended in the boiler by an upward stream of hot air. The coal is burned in this ebullient, liquid-like mixture, hence the name "fluidized." As the coal burns, sulfur gases from coal combine with limestone to form a solid compound that is recovered with ash.

Fly Ash
The finely divided, inert particles of ash in flue gases arising from the combustion of fuel.

Fossil Fuel
Any naturally occurring fuel of organic nature including coal, crude oil, natural gas, and others.

Gob
The term generally applied to that part of the mine from which the coal has been removed and partially filled with waste.

GPS
A Global Positioning System uses signals from satellites to pinpoint the exact location of objects such as mobile mining equipment.

Ground Control
Measures taken to prevent roof falls or coal bursting from the seam due to pressure or other factors.

Haulageway
An underground entry or passageway that is designed for the horizontal transport of coal, supplies, or waste via track or belt conveyor (vertical transport is called hoisting).

Heaving
The rise of the mine floor due to pressure from coal pillar supports left after most of the coal has been removed.

High Sulfur Coal
This is coal which naturally contains a large amount of sulfur that converts into sulfur dioxide upon burning.

Highwall Mining
A highwall mining system consisting of a remotely controlled, continuous mining machine which extracts coal from a horizontal cut in a highwall bench (which may reach depths of several hundred feet or deeper). The coal is conveyed via augers, belt, or chain conveyors to the outside.

Hogsback
A sharp rise in the floor of a seam of coal.

Horseback
A mass of material shaped like a horse's back located in the roof and covered by a slippery surface.

Hydraulic
Of or pertaining to fluids in motion. These fluids are often compressed by pumps and are used to mechanically control various parts of machines used in the mines such as jacks, cutters, loaders, and transport vehicles.

Hydro Seeder
A machine which sprays a blue-green liquid containing grass seed, fertilizer, and a binding material on topsoil to prevent erosion until the grass seed can sprout and take root.

Hyperbaric Filtration
Removing water from coal by applying extreme pressure.

IGCC
Advanced Fuel Technology—A technical name which represents a gasification process known as "Integrated Gasification Combined Cycle." This is a partnership which brings together the Commonwealth of Kentucky, The Kentucky Coal Council, the University of Kentucky Center For Applied Energy Research, and Global Energy, Inc. to provide low cost electricity and fuel to be burned in transportation vehicles by using Kentucky coal and solid waste generated by private citizens and business and industry.

In Situ
In the natural or original position. Applied to rock, soil, or fossils when found as they were originally formed or deposited.

Intake
The passage through which fresh air is drawn or forced into a particular area of the mine.

Kerf
The undercut of a coal face.

Kettle Bottom
A smooth, rounded piece of rock, cylindrical in shape, which may drop out of a mine roof without warning. Its origin is thought to be the remains of the stump of a tree which has been replaced by sediments so the original form has been preserved.

LEBS
Low Emission Boiler System developed by the National Energy Technology Laboratory as a highly efficient, advanced power system with a significantly high thermal efficiency, and superior environmental performance which produces a lower cost for electricity.

Lignite
The softest coal with the highest moisture and lowest BTU content. Lignite is being used more and more for generating electricity in certain areas of the United States and is also being converted to synthetic gas.

Liquefaction
Converting coal into a synthetic liquid fuel similar in nature to crude oil and /or refined products such as gasoline.

Loader or Loading Machine
Any device used to transfer broken or blasted coal or other material into haulage equipment.

Longwall
A system of mining coal in which all of the coal is removed except the shaft pillars and sometimes the main road pillars. This mining usually occurs on the long side of a block of coal with a complex machine containing rotary cutting drums that are usually pulled along the face by a gear driven chain which shears the coal from the face. The machine usually contains its own roof support system consisting of large diameter hydraulic jacks and thick metal plates which are fitted against the top and floor of the mine.

Magazine
A structure used for storing explosives.

Main Entry
This is one of a main group of entries which proceed underground through the coal.

Main Fan
A mechanical device containing a large electric motor and fan blades which ventilates by exhausting or blowing to induce airflow through the coal mine's roadways and workings.

Man Door
A door installed in a permanent stopping to allow personnel to travel from one entry to another. The door must be constructed so as to prevent air from leaking through the stopping when closed.

Mantrip
A device, usually a rail or rubber tire car, used to transport men to and from various work areas in a coal mine.

MESA
Mining Enforcement & Safety Administration. In 1973, the Secretary of the Interior created a new departmental agency separate from the Bureau of Mines which assumed safety and health enforcement functions while the Bureau retained the functions of mineral resource development. MESA was the predecessor to MSHA (Mine Safety and Health Administration).

Metacarpal Gloves
Special gloves designed to protect the backs of a miner's hands.

Methane CH_4

A potentially explosive gas formed naturally from the decay of vegetative matter commonly found in coal mines. Its properties include being odorless, colorless, tasteless, and nontoxic. Methane mixtures are often called "firedamp."

Methane Monitor

An electronic instrument which indicates and measures the presence of methane. These are usually mounted on a piece of mining equipment.

Micronized Coal

Coal which is pulverized to the consistency of talcum powder. This material can be combined with micronized limestone to capture sulfur dioxide emissions when coal is burned. Nitrogen oxide emissions can also be reduced through this clean coal technology process.

Mine Foreman

A certified person placed in charge of the workings of the mine and of the persons employed there.

Mineral Processing

Any method of cleaning coal of undesirable content.

Mine Roof

The layer of hardened clay, limestone, sandstone, or other material that lies over the coal bed or seam.

Mine Run Coal

Unscreened coal, just as it is mined, less dirt and slate.

Mine Ventilation

The process of providing a directed flow of fresh and return air along all underground roadways, manways, and areas where work is occurring in a coal mine.

Mountain Break

Deep fissures created from settling in the earth's strata.

MRE Instrument

A gravimetric dust sampler with a four channel elutriator. This measures the presence of dust and was developed by the Mining Research Establishment of the National Coal Board of London, England.

MSHA

Mine Safety and Health Administration. This is the federal agency which regulates coal mine health and safety.

MVM

Mine Ventilation Manager was developed by the University of Kentucky to perform advisory consultations for underground environmental emergencies.

NETL

The National Energy Technology Laboratory located in Pennsylvania and West Virginia is a federally owned and operated research and development entity charged with solving national energy and environmental problems.

Nips

Devices at the end of a trailing cable which are used for connecting the cable to a power source (usually a trolley wire and ground).

Nitrogen Oxide Emissions
Nitrogen oxide is released when coal is burned. When combined with water, nitric acid is formed.

Open-Pit Mining
In this surface mining system, huge powered shovels and trucks remove the overburden in levels called benches. Shovels or draglines are used to remove and load coal into trucks. This method is most commonly used when the coal beds are extremely thick. After the coal is removed, the mountain tops are reclaimed either as flat or rolling contours. The rock from the mountain top is placed in hollow fills—very similar to fills along the construction of highways.

OSM
Office of Surface Mining Reclamation and Enforcement.

Outcrop Coal
Coal found at the outer perimeters of the seam and at or near the surface. It is usually, but not always, considered to be of inferior quality due to oxidation.

Overburden
The material of any nature (usually limestone, sandstone, clay, dirt, etc.) that overlies a deposit.

Overcast or Undercast
An enclosed airway which permits one air current to pass over or under another without interruption.

Panel
A block of coal which comprises one mineable unit.

Parting
A small joint in a coal seam or a layer of rock in a coal seam.

PCD
Particulate Control Devices are used to clean hot gases.

Peat
The partially decayed plant matter found in swamps and bogs. This is one of the earliest stages of coal formation.

Penetrometer
Instrument used to measure the strength and deformities in rock.

Permissible
Anything that is allowable or permitted by law. It is most widely applied to mine equipment or explosives. These items have passed certain tests and are determined to be safe under specified conditions where hazards from explosive gas or coal dust exist.

Permit
A document issued by a state and/or federal regulatory agency that gives approval for mining operations to take place.

Pillar
A block of coal left to support the overlying strata in a mine. These are sometimes left as permanent support for the roof of a coal mine.

Pinning
Roof bolting.

Pneumoconiosis (Black Lung)
A chronic respiratory disease arising from breathing coal dust. Also referred to as "Black Lung."

Portal
The immediate entrance to a mine. Usually the mouth of the tunnel at the mine.

Post-Combustion Cleaning
Cleaning emissions from burning coal between the boiler and the smokestack.

Powered Shovel
A large machine with a boom arm and bucket powered by electricity or diesel hydraulics used to dig and remove overburden in the surface mining process.

Pre-Combustion Cleaning
When sulfur and other mineral matter is removed from coal prior to combustion, sulfur dioxide emissions are reduced in combustion gases.

Preparation Plant
A coal processing facility designed for crushing, sizing, and washing coal to prepare it for use by customers.

Pressurized Fluidized Bed Combustion
A clean coal technology that is similar to atmospheric fluidized bed combustion except the boiler is pressurized up to 16 times the normal atmospheric pressure. Sulfur dioxide emissions are reduced by 95 percent and nitrogen oxide is reduced moderately.

Primacy
A term which is used in assigning responsibility for regulation. This law gives states primary responsibility for regulation and the federal government a supervisory or oversight role.

Proximate Analysis
A physical, non-chemical test of the constitution of coal, useful for determining the commercial value of coal and basically using a one gram sample of coal which is heated under controlled conditions at fixed temperatures and times to determine moisture, volatile matter, fixed carbon, sulfur, ash content, and BTU.

PSDF
The Power Systems Development Facility is a co-partnered project with private industry and the U.S. Department of Energy charged with testing new process configurations for advanced coal-based power generating systems.

Reclamation
The restoration of land and environmental values to a surface mine after the coal has been extracted. These operations begin as soon as the coal is removed from the mine site. This process includes restoring the land to its approximate original appearance by replacing the topsoil and planting native trees, grasses, and ground covers.

Reforestation Initiative
A reclamation initiative which involves leaving rocks and dirt in a less dense state with the theory that the roots of trees and other vegetative matter can become established more quickly.

Refuse
Extra fine coal, water, dirt, and rock which results from washing coal at a preparation plant.

Reserves
The portions of the identified coal resource that can be economically mined at the time of determination. The reserve is derived by applying a recovery factor to that component of the identified coal resources designated as the reserve base.

Resin Bolting
A newer method of placing roof bolts in the strata over the coal seam to provide roof control. This method employs a striated steel rod grouted with an epoxy resin which glues it to the surfaces through which it passes.

Retreat Mining
Coal extraction which occurs in the opposite direction of development or advance mining. Retreat usually begins when the entries have reached the property boundaries and the coal company begins to remove certain support pillars as they withdraw toward the mine entrance.

Return Air
The air which has passed through all the working faces and is being returned to the outside.

Rib
The side of a pillar of coal or wall of an entry. The solid coal on the side of any underground passage within the mine.

Rider
A thin seam of coal overlying a thicker one.

Robbing Pillars
The systematic removal of coal pillars between rooms or chambers to regulate the subsidence of the roof. Also called bridging back the pillar, drawing the pillar, or robbing the pillar.

Rock Dust
A white dust made of powdered limestone which is applied to the roof, ribs, and floor of an underground mine to contain or minimize explosions. This material also improves the reflected light within the mine.

Rock Mechanics
The characteristics and nature of rock.

Roof
The stratum overlying a coal seam. The overhead surface of an underground coal mine.

Roof Bolt
A long, steel bolt inserted in a drilled hole which is used to support the mine roof. A unit of this device consists of a bolt, a steel bearing plate, and sometimes an expansion shell.

Roof Fall
Material from the roof which falls to the floor of the coal mine.

Roof Stress
When coal is extracted the internal forces become unbalanced and pressures are exerted on the roof of the mine.

Room & Pillar Mining
A system of mining in which approximately half of the coal is mined by advancing rooms off the entries and leaving narrow pillars in between. The pillars are then retrieved when the boundaries are reached and retreat mining toward the mine entrance begins.

Safety Fuse
A train of powder enclosed in cotton, jute, or other water proofing compounds which burns at a uniform rate and is used to fire a cap containing a detonating compound which, in turn, sets off the explosive charge.

Sandstone
A sedimentary rock consisting of quartz sand united or cemented by materials such as iron oxide and/or calcium carbonate.

Scoop
A rubber tire, battery or diesel powered piece of equipment designed for cleaning roadways and hauling supplies.

Scrubber
Any of several forms of chemical/physical devices that operate to remove sulfur compounds from coal during combustion. These devices combine the sulfur in gaseous emissions with another chemical compound medium to form inert sludge, which is removed for disposal or sold as a by-product.

Seam
A stratum or bed of coal

SCSR
Self Contained Self Rescuer. A self-contained one-hour supply of oxygen worn by individual coal miners.

Self-Rescuer (Filter Type)
Small filtering device using a canister and attached rubber mouthpiece carried by an underground miner, either on a belt or in a pocket, to provide immediate protection against carbon monoxide and smoke in case of a mine fire or explosion. The canister contains a layer of fused calcium chloride that absorbs water vapor from the mine air. The wearer breathes through the mouth, the nose being closed by a clip. This device is used for escape purposes only, because it does not sustain life in atmospheres containing deficient oxygen. The length of time a self-rescuer can be used is governed mainly by the humidity in the mine air, but it is usually between 30 minutes and one hour.

Shaft
A vertical opening through the strata that connects the surface with the underground workings. The shaft may be used in connection with the mining of coal, for the purpose of ventilation or drainage, or for hoisting personnel, coal, or other materials.

Shaker Conveyor
An outmoded transport device equipped with metal troughs linked together where coal is moved forward by a shaking or sliding motion created by an eccentric motion arm which thrusts the trough forward and then back once each cycle.

Shale

A rock formed by consolidation of clay, mud, or silt which usually has a laminated structure of layers loosely bound together.

Shear or Shearer

A mining machine primarily used for longwall faces that uses a larger diameter, relatively narrow width, rotating head of bits which shear the coal as it progress along the face.

Shields

Large heavy gauge metal boxes mounted overhead on a longwall mining machine which are supported by hydraulic jacks to hold the roof and protect the miner from falling debris.

Shortwall

A system of mining, with a much shorter face than longwall, which uses a continuous miner rather than a shear.

Shuttle Car

A trackless, one unit, self-powered haulage and self-discharging device used for short distances to transport broken coal from the loading machine to the conveyor.

Slack

Small coal which is usually less than one inch in diameter.

Slag

The waste product from the process of melting.

Slate

A miner's term for any shale or slate which is found in or around coal. Geologically, it is a dense, fine textured, metamorphic rock, which has excellent paralleled cleavage so that it breaks into thin plates or pencil-like shapes.

Slope

An inclined tunnel leading to a coal seam.

SMCRA

The Surface Mining Control And Reclamation Act of 1977 which created national environmental standards for coal mining.

Smoke Roller

A special roller mounted on a conveyor belt which acts as a safety cutoff if the load becomes too great for the drive motor.

Spad

A flat spike which is hammered into a wooden plug then placed in a hole drilled in the mine roof. A plumb line can be threaded through the end. The attached plumb bob assists in establishing a particular point much the way a stake is used in surveying on the surface.

Squeeze
The settling, without breaking, of the mine roof. This causes a gradual upheaval of the mine floor due to the weight of the overlying strata.

Steel-Toed or Safety-Toed Boots
Mining boots equipped with a steel plate across the toe for protection from injury.

Steam Coal
Coal which is usually lower in BTU content than metallurgical grade and is used in a power plant or industrial steam boiler to produce electricity.

Stemming
The noncombustible material used on top or in front of a charge or explosive.

Stopping
Any wall built to deflect or impede the flow of ventilating air.

Stratum
A sheet-like layer of sedimentary rock.

Stump
Any small pillar of coal.

Sub-Bituminous Coal
A coal with a heating value between bituminous and lignite and with low-fixed percentages of volatile matter and moisture. Sometimes referred to as black lignite.

Subsidence
A sinking or settling of the ground surface caused by the collapse of the underground mine roof in mined out areas.

Sulfur Dioxide Emission
Coal contains sulfur which converts to a gas upon burning. The sulfur dioxide gas combines with atmospheric water to form sulfuric acid/sulfate.

Surface Mine
Area of land from which coal is extracted by surface methods.

Tail Piece
The pulley or roller in the tail or foot section of a belt conveyor around which the belt runs. This is also known as a foot section pulley.

Timber
A collective term for all underground wood supports.

Tipple
Originally the place where mine cars were tipped and emptied of their coal. Currently this applies to several surface structures of a mine including the preparation plant and load-out facilities where coal is shipped.

Tram
This is a general term referring to the moving of self-propelled equipment.

Tree Cookies
A wafer of wood cut from the trunk of a tree which is used to analyze growth patterns.

Trip
A train of coal cars.

Tunnel
A horizontal (sometimes with a slight incline) underground passageway, entry, or haulage way that is open to the surface at both ends. A tunnel must pass completely through the hill or mountain.

Ultimate Analysis
Precise determination, by chemical means, of the elements and compounds contained in coal.

Undercut
To cut below or undermine the coal face by removing a portion of the bottom of the seam with a pick or cutting machine.

Unit Train
A long rail train of between 60 and 150 or more hopper cars which carry coal between a mine and a destination.

Vacuum Filtration
A process of removing water from coal via a vacuum process.

WCPM
Work Crew Performance Model was developed by the University of Kentucky and is a practical way to make sure safety is paramount during the performance of routine jobs which involve high-risk tasks.

White Damp
CO or carbon monoxide. This is a gas that can be present in the afterdamp of a gas or dust explosion or in gases given off by a mine fire. Carbon monoxide is also one of the constituents of the gases produced by blasting. CO is absorbed by the hemoglobin of the blood to the exclusion of oxygen. One tenth of 1% may be fatal in 10 minutes.

Working Face
Any place in an underground mine where material is extracted from its natural deposit during a mining cycle.

Working Section
The face of the coal to the point where it is being loaded on conveyors, rail cars, or other haulage devices to begin its trip to the outside of the underground mine.

Sources for glossary include:

Illinois Clean Coal Institute. *Clean Coal Technology (CCT) Glossary.* Carbondale, Illinois: The Illinois Department of Commerce and Community Affairs, Illinois Clean Coal Institute, Available: http://www.icci.org/glossary.html [1999, January 14].

Kentucky Mining Institute. *Coal Mining Reference Book.* Lexington, Kentucky: Kentucky Mining Institute, 1997.

United States Department of Labor. *Judges' Benchbook of The Black Lung Benefits Act.* Washington, D.C.: United States Department of Labor Office of Administrative Law Judges, 1997.

BIBLIOGRAPHY

Books

Adler, Irving and Ruth. *Coal*. New York: John Day Co. Inc., 1965.

Arnold, Guy. *Coal*. New York: Gloucester Press, 1985.

Berkowitz, Norbert. *An Introduction to Coal Technology*. New York: Academic Press, 1994.

Brestensky, Dennis F., Evelyn A. Hovanec, and Albert N. Skomra. *Patchwork Voices: The Culture and Love of a Mining People*. Pittsburgh, PA: University of Pittsburgh Press, 1991.

Burt, Olive. *Black Sunshine: The Story of Coal*. New York: Julian Messner, 1977.

Campbell, Shirley Young. *Coal & Coal People*. Parsons, WVa: McClain Printing Company, 1994.

Chaffin, Lillie D. *Coal Energy and Crisis*. NY: Harvey House, 1974.

Davey, John. *Coal Mining*. London: A&C Black, Ltd., 1960.

Davis, Bertha. *The Coal Question*. New York: F. Watts, 1982.

Dionetti, Michelle. *Coal Mine Peaches*. New York: Orchard Books, 1991.

Dix, Keith. *What's A Coal Miner to Do? The Mechanization of Coal Mining*. Pittsburgh, PA: Univ. of Pittsburgh Press, 1988.

Doty, Roy. *Where Are You Going With That Coal?* New York: Doubleday & Co., Inc., 1977.

Fay, Albert H. *A Glossary of the Mining and Mineral Industry*. Lincoln, NE: University of Nebraska Press, 1979.

Francis, Wilfred. *Coal: Its Formation and Composition*. London: E. Arnold, 1961.

Goode, James B. *Ancient Sunshine: The Story of Coal*. Ashland, KY: Jesse Stuart Foundation, 1997.

Goode, James B. *Up From The Mines*. Ashland, KY: Jesse Stuart Foundation, 1993.

Graebner, William. *Coal Mining Safety in the Progressive Period*. Lexington, KY: University Press of Kentucky, 1976.

Green, Archie. *Only A Miner: Studies in Recorded Coal Mining Songs*. Urbana, IL: University of Illinois Press, 1972.

Harter, Walter L. *Coal: The Rock That Burns*. New York: E. P. Dutton, 1979.

Harvey, Curtis E. *Coal In Appalachia*. Lexington, KY: The University Press of Kentucky, 1986.

Harvey, Curtis E. *The Economics of Kentucky Coal*. Lexington, KY: The University Press of Kentucky, 1977.

Hossner, Lloyd R., Editor. *Reclamation of Surface-Mined Lands*. CRC Press, 1988.

Illinois Clean Coal Institute. *Clean Coal Technology (CCT) Glossary*. Carterville, IL. Available: <http://www.icci.org/glossary.html>

Kamecke, Theo. *Coal: Bridge to the Future*. Woodbury, New York: J.N. Company, 1980.

Kentucky Coal Council and The Kentucky Coal Association. *"Kentucky Coal Facts: 1997-1998 PocketGuide."* Lexington, KY, 1998.

Kentucky Mining Institute. *A Coal Mining Reference Book, Fifth Edition, Revised*. Lexington, KY, 1997.

Khoury, D. L., Editor. *Coal Cleaning Technology*. Noyes Publications, 1981.

Lombardi, John and Charles V. Jude. *Mechanical Excavation Systems*. United States Department of the Interior, 1996.

Mitgutsch, Ali. *From Swamp to Coal*. Minneapolis, MN: Carolrhoda Books, 1985.

Naylor, Richard and Lorrie Smith, Editors. *Directory of Coal Information Sources for Kentucky*. Coal Information Network, 1984.

Rogers, William Patrick. *The Coal Primer: A Reference Handbook For Non-Coal People*. Little Rock, AR: William P. Rogers, 1978.

Schobert, Harold H. *Coal: The Energy Source of the Past and Future*. American Chemicals Society, 1987.

Theiss, Nancy. *King Coal in Kentucky*. Louisville, KY: The Courier Journal & Louisville Times, 1985.

Thomas, Larry. *Handbook of Practical Coal Geology*. New York: John Wiley & Sons, 1992.

Todd, A.H.J. *Lexicon of Terms Relating to the Assessment and Classification of Coal Resources*. Crane Russak & Co., 1983.

United States Department of Labor. *Judges' Benchbook of The Black Lung Benefits Act: January 1997*.

Wagoner, E.L., et al. *Coal: Current Advances in Coal Chemistry and Mining Techniques*. Nettleton, MA: Ayer Company Publishers, 1976.

Video & Multimedia:

American Coal Foundation. 1130 Seventeenth Street, NW, Suite 220, Washington, D.C. 20036-4604. *Coal Today* (Video).

American Electric Power Service Corp. Public Affairs Office. PO Box 700, Lancaster, OH 43130. *All About Coal* (Video).

American Electric Power Service Corporation. Public Affairs Office. PO Box 700, Lancaster, OH 43130. *Coal: The Inside Story* (Video).

BHP Minerals. 300 W. Arrington, Suite 200, Farmington, NM 87401 *Coal—Born in the USA* (Video)

Caterpillar, Inc. SME Headquarters, PO Box 625002, Littleton, CO 80162-5002. *Common Ground.* (Video). Nashville, TN: Capital Communications, 1993.

Center for Energy and Economic Development. 1800 Diagonal Road, Suite 370, Alexandria, VA 22314. *America's Fuel* (Video).

Coal Power (Video). PO Box 6608, Bozeman, MT 59715. Available: <http:www.coalpower.com>

Goode, James B. *Coal, Steel, Machines, & Men: The Benham Story* (Video). Benham, KY: The University of Kentucky Southeast Community College, 1993.

Goode, James B. *Lynch: A Coal Legacy* (Video). Benham, KY: The University of Kentucky Southeast Community College, 1990.

Lignite Energy Council. PO Box 2277, Bismarck, ND 58502). *Electrical Generation from North Dakota Lignite* (Video). *The Environment and Lignite Energy* (Video). *Lignite Energy: It's Good for All of Us* (Video). *Lignite Mining and Reclamation* (Video).

National Mining Association. Mining: Discoveries For Progress. *Tell Everyone About Mining* (Video). 1130 Seventeenth Street, NW, Washington, DC 20036-4604: American Mining Congress, 1993.

Peabody Group. Public Relations Department. 701 Market St., Suite 825, St. Louis, MO 63101. *An Enduring Legacy: Reclaimed Lands on the Black Mesa* (Video). *Coal People: Century of Pride* (Video). *One Future: Safety Through Teamwork* (Video). *Partnerships: A Story of Coal from the Black Mesa* (Video).

Western Kentucky University, KET, The Kentucky Network, and Kentucky Coal Council. *Kentucky Coal: A Multimedia Education Kit. Coal: Kentucky's Power Source (Video).* Lexington, KY, 1997.

Web Sites, Magazines, & Newspapers:

Clean Coal Today. Germantown, MD 20874: U.S. Department of Energy FE—22,270CC.

Coal Age. 29 North Wacker Avenue, Chicago, IL 60606. Available: <http://www.coalage.com>.

Coal Daily. 1800 Massachusetts Ave., NW, Suite 500, Washington, D.C. 20036. Available: <http://www.fieldston.com/cd.html>

Coal Education. <http:coaleducation.org>

The Coal Journal. PO Box 3068, Pikeville, KY 41502.

The Coal Miner's Exchange. PO Box 1441, Rt. 61 North, Oak Hill, WV 25901.

Coal Outlook. Pasha Publications, Inc., 1600 Wilson Blvd., #600, Arlington, VA 22209. Available: <http://www.ftenergyusa.com/coaloutlook/default.asp>

Coal People. Al Skinner Enterprises, Inc., PO Box 6247, Charleston, WV 25362.

Coal and Synfuels Technology. 1616 N. Fort Meyer Drive, Suite. 1000, Arlington, VA 22209.

Coal Transportation Report. 1800 Massachusetts Ave., N.W., Suite 500, Washington, D.C. 20036. <http://www.fieldston.com/ctr.html>

Engineering and Mining Journal. 29 North Wacker Drive, Chicago, IL 60606.

Fossil Energy Review. U.S. Department of Energy, Office of Communications, Fossil Energy, FE-5 Washington, D.C. 20585.

Green Lands. West Virginia Mining & Reclamation Association, 1624 Kanawha Boulevard East, Charleston, WV 25311.

The Holmes Safety Association Bulletin. Mine Safety and Health Administration-US DOL, 5th Floor, EPD #535A, 4015 Wilson Blvd., Arlington, VA 22203-1984.

International Journal of Coal Geology. Southern Illinois University, Department of Geology, Carbondale, IL 62901. Available: <http://www.elsevier.nl/inca/publications/store/5/0/3/3/7/>

Morrison, George. *4 Your Info: The Rock That Burns: Kentucky Coal Keeps the Lights Glowing, the Economy Moving*. Special to the Louisville Courier Journal, Newspaper in Education.

United States Department of Energy-Energy Information Administration. *Electric Power Annual*, Vol.1, 1997. Available: <http://www.eia.doe.gov/pub/electricity/epav196.pdf>

U.S. Geological Survey: Energy Resource Surveys Program. *Assessing The Coal Resources of the United States*. USGS Fact Sheet FS-157-96: July, 1996.

Vaughn, Robert. *A Guide to Appalachian Coal Hauling Railroads*. <http://w.spikesys.com/trains/app_coal.html>

Organizations:

American Coal Foundation. 1130 17th Street, N.W., Suite 220, Washington, D.C. 20036-4604 Available:<http://www.acf-coal.org/>

American Institute of Mining, Metallurgical, and Petroleum Engineers. New York, NY <http://www.idis.com/aime/>

Center for Energy and Economic Development. 1800 Diagonal Road, Suite 370, Alexandria, VA 22314. Available: <http://www.ceednet.org/>

Kentucky Coal Association. 340 South Broadway, Suite 100, Lexington, KY 40508. Available: <www.kentuckycoal.org>

Kentucky Coal Council. 709 Millpond Road, Lexington, KY 40514. Available: <www.coaleducation.org/kcmec>

Kentucky Geological Survey. 228 Mining and Minerals Resources Bldg., University of Kentucky, Lexington, KY. 40506 Available: <www.uky.edu/kgs/home.htm>

KET, The Kentucky Network. Lexington, KY. Available: <www.ket.org/Trips/coal/index.htm>

National Energy Information Administration. Washington, D.C. Available: <http:www.eia.doe.gov/fuelcom/.html>

National Mining Association. 1130 17th Street, N.W., Washington, D.C. 20036 Available: <http://www.nma.org>

University of Kentucky Center for Applied Energy Research. 2540 Research Park Drive, Lexington, KY. Available: <www.caer.uky.edu>

SPECIAL THANKS TO:

Larry Adams, Vice President
Mining Consulting Services, Inc.

American Commercial Barge Lines

Arch of Kentucky

Bill Caylor, President
Kentucky Coal Association

Charolais Coal Corporation

CONSOL Coal Group

Continental Conveyor & Equipment Company

Tim Cox, Photographer

CSX Transportation

Cyprus Mountain Coals

Doug Deak
National Energy Technology Laboratory

Susan and David Duff
Pine Branch Coal Sales

East Kentucky Power Cooperative

"J" Hamon
Kentucky Department for Surface Mining Reclamation and Enforcement

Fairchild International

Steve Gardner, President
Mining Consulting Services, Inc.

Jamie Goode, Photographer
Illinois State University

J. Dan Guffey, P.E., P.L.S.
Coal Development Projects
Kentucky Coal Council

Heavy Machines, Inc.

Hitachi

Elmer Howard

Ingersol-Rand Construction & Mining

Innovative Mining & Equipment, L.L.C.

Jesse Stuart Foundation

Julie Ross, Geologist
Mining Consulting Services, Inc.

J. Eric King,
Senior Engineering Technician
Mining Consulting Services, Inc.

Kentucky Coal Association

Kentucky Coal Council

Komatsu America International Co.

Aleck Leedy, Research Assistant
University of Kentucky
Mining Engineering Department

Joseph E. Lema, Vice President
Manufacturers and Services Division
National Mining Association

Wendong Li, Research Assistant
University of Kentucky
Mining Engineering Department

Lodestar Energy

Long-Airdox Virginia Tech
Corporate Research Center

Bill Marcum, Vice President
Kentucky Coal Association

Dr. Thomas Marrero
University of Missouri Columbia and Rolla Campuses

Martin County Coal Corporation

Mike Musulin II

Peabody Wildlife Management Area

Steven Rapchak, Chairman
Mining Consulting Services, Inc.

Rudd Equipment Company

Susan Shaw
Mining Consulting Services, Inc.

Donna J. Slone
Evening Star Communications

Dr. Joe Sottile, Associate Professor
University of Kentucky
Mining Engineering Department

Jason Stoltz, Research Assistant
The University of Kentucky
Mining Engineering Department

Dr. Richard J. Sweigard, Chair
University of Kentucky
Mining Engineering Department

Dr. Daniel Tao, Assistant Professor
University of Kentucky
Mining Engineering Department

Terex Mining

United States Department of the Interior

Dr. John W. Wilson
University of Missouri Columbia and Rolla Campuses

Dr. Andrew Wala, Professor
University of Kentucky
Mining Engineering Department

David Zegeer, Mining Engineer